Beginning Arduino ov7670 Camera Development

Robert Chin

Copyright © 2015 Robert Chin

All rights reserved.

Copyright © 2015 Robert Chin

All rights reserved.

Table of Contents

About the Author iv

Introduction v

Chapter 1: Introducing the Omnivision OV7670 Camera 1

Chapter 2: Introducing the Arduino 19

Chapter 3: Arduino Programming Language Basics 39

Chapter 4: Digital Design Review 53

Chapter 5: Taking Photos with the Omnivision ov7670 Camera – Part 1 77

Chapter 6: Taking Photos with the Omnivision ov7670 Camera – Part 2 185

Appendix A: Camera Register Defines 225

Appendix B: Image Capture Program Variables 238

About the Author:

Robert Chin has a Bachelor of Science degree in computer engineering and is experienced in Arduino camera development, C/C++, Unreal Script, Java, DirectX, OpenGL, and OpenGL ES 2.0. He has written 3d games for the Windows, and Android platforms. He is the author of "Beginning Android 3d Game Development", and "Beginning IOS 3d Unreal Games Development" both published by Apress and was the technical reviewer for "UDK Game Development" published by Course Technology CENGAGE Learning.

Introduction

This book is meant to be a quick start guide to using the Omnivision ov7670 digital camera. I show you in a detailed step by step hands on example how to take photos with the ov7670 camera using an Arduino Mega 2560 and how to use a SD card reader/writer to save these images to an SD card. Then, I show you how to transfer these images to your computer and convert them to a common image format that is easily viewable. This book would also be beneficial to those that want to develop Arduino programs for cameras other than the ov7670 since much of the information presented here can also be applied to other digital cameras.

Note: Chapter 6: Taking Photos with the Omnivision ov7670 Camera – Part 2 is the chapter you want to read if you want to quickly put together a working camera system.

A summary of the content of the book's chapters follows.

Chapter 1: "Introducing the Omnivision OV7670 Camera" – In this chapter I discuss in detail the ov7670 camera in terms of its features, operation, and steps needed in order to take a photo.

Chapter 2: "Introducing the Arduino" – Here I start by giving some background information about the Arduino. I then give an in depth discussion of the Arduino Mega 2560 board, and then I guide the reader through a hands on example where I show you how to setup your Arduino and how to get a simple program that controls the blinking of a light working.

Chapter 3: "Arduino Programming Language Basics" – In this chapter I discuss the basics of the Arduino programming language.

Chapter 4: "Digital Design Review" – In this chapter I cover how the ov7670 works at a chip level. I cover the main camera chip, the frame buffer memory chip, and how these two chips are connected together. I discuss the general procedure to capture a video frame to the camera's frame buffer memory and how to read the image data from the camera's frame buffer memory.

Chapter 5: "Taking Photos with the Omnivision ov7670 Camera – Part 1" – In this chapter I discuss the SD card reader/writer, the I2C interface, and the Arduino program or "sketch" I wrote to capture an image from the camera and then to save it on a SD card. I also cover ffmpeg which is used to convert the images produced by the camera into common easily viewable images.

Chapter 6: "Taking Photos with the Omnivision ov7670 Camera – Part 2" – In this chapter I present a hands on example where I show you step by step how to take a photo with the camera and save it to an SD card using the image capture software I developed. This chapter explains everything from the connections required, how to use the image capture software, and how to convert the final images to an easily viewable format. This is the chapter you want to read if you just want to quickly put together a working camera system.

Appendix A: "Camera Register Defines" – This appendix lists all the important camera registers and values

Appendix B: "Image Capture Program Variables" – This appendix lists all the variables that were used in my image capture program.

Chapter 1

Introducing the Omnivision OV7670 Camera

In this chapter I cover the Omnivision ov7670 camera. First, a short description of the camera is given followed by some photos of the camera itself. Then key digital camera terminology needed to understand key concepts in this book are covered. I then give a more in depth explanation of the camera including details of each key part and the steps by which an image is captured, processed and transmitted to the Arduino.

What is the OV7670 Camera?

The ov7670 camera is a low cost widely available CMOS camera made by Omnivision Technologies located in Santa Clara, California. It comes in two versions one without frame buffer memory and one with frame buffer memory which is commonly called the FIFO version. In this book we will use the version with the FIFO frame buffer memory. The frame buffer memory holds image data that has been captured from the camera. The image data can then be transferred from the frame buffer memory to the Arduino's memory or to a storage device such as a SD Card. The ov7670 camera can be used with the Arduino through its SCCB interface that is compatible with the Arduino's I2C interface. The camera can be focused manually by turning the camera lens clockwise and counterclockwise which moves the lens outward and inward. The maximum resolution of the camera is VGA which is 640 pixels wide by 480 pixels high. Figure 1-1 shows a photo of the back side of an ov7670 camera with frame buffer memory labelled "Averlogic". Figure 1-2 shows a photo of the front of an ov7670 camera. Figure 1-3 shows a picture captured from a ov7670 camera.

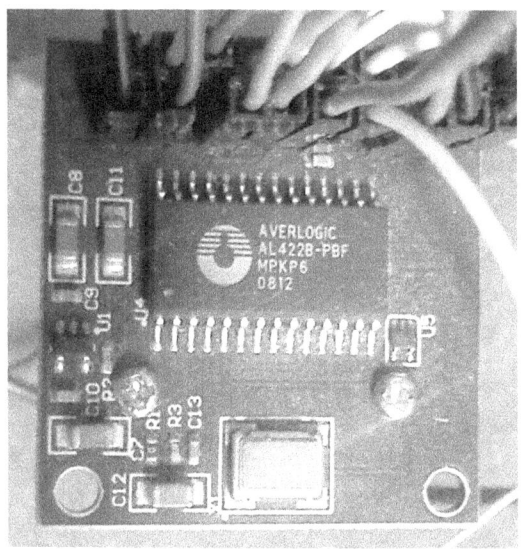

Figure 1-1. ov7670 FIFO camera version back side showing the Averlogic frame buffer memory

Figure 1-2. ov7670 FIFO camera version front side showing camera lens

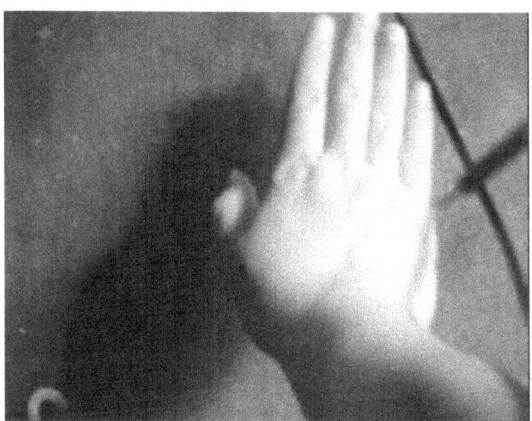

Figure 1-3. Picture captured from an ov7670 camera

Key Camera Terminology

This section covers key terms related to digital cameras and traditional cameras.

- Pixel – A pixel is the smallest unit that makes up a digital image. It is generally a small square illuminated element that can take on various colors. For example, in Figure 1-4 you can see that the number "0" is composed of many pixels represented in the figure by black squares.

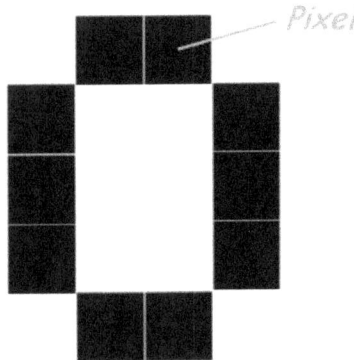

Figure 1-4. Group of pixels representing the image of the letter "O"

- Resolution – Resolution refers to the width and height of an image in pixels.

- VGA – VGA refers to a camera resolution that generates images that are 640 pixels wide and 480 pixels high.

- QVGA – QVGA refers to a camera resolution that generates images that are 320 pixels wide and 240 pixels high.

- QQVGA – QQVGA refers to a camera resolution that generates images that are 160 pixels wide and 120 pixels high. See Figure 1-5 for a comparison of the VGA, QVGA, and QQVGA resolutions.

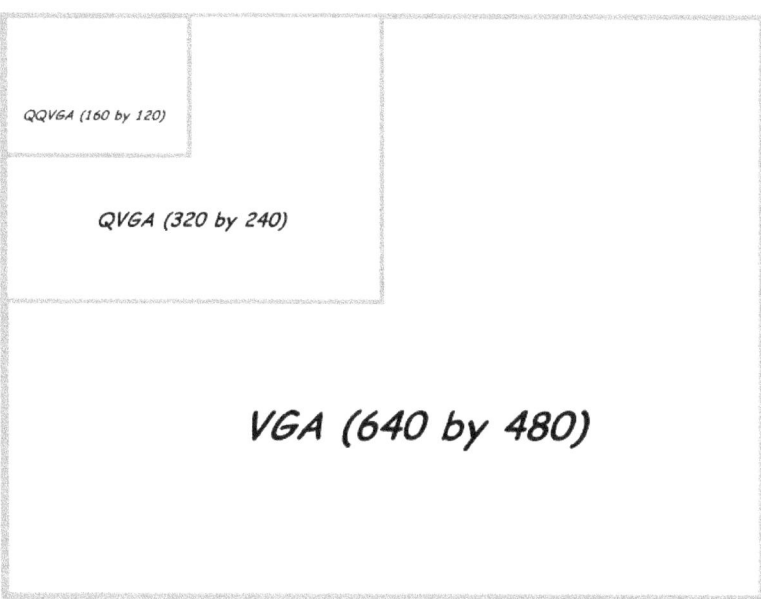

Figure 1-5. Comparison of VGA, QVGA, and QQVGA resolutions

- CIF – CIF refers to a camera resolution that generates an image that is 352 pixels wide and 288 pixels high. See Figure 1-6 for a comparison with the VGA modes.

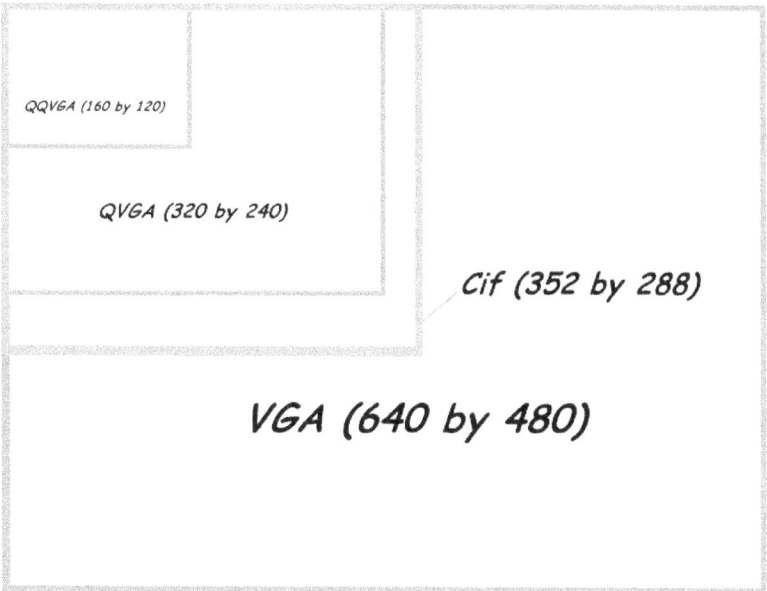

Figure 1-6. Comparison of CIF with VGA, QVGA, and QQVGA resolutions

- YUV –YUV is an image encoding method and is discussed in more detail later in this book. The pixels that make up a digital image must be represented internally in an image format and YUV is one of the available image formats used to represent these pixels. The YUV format incorporates luminance or brightness values and color values.

- YCbCr – YCbCr is an image encoding method and is discussed in more detail later in this book. The pixels that make up a digital image must be represented internally in an image format and YCbCr is one of the available image formats used to represent these pixels. In the YCbCr image format a pixel incorporates luminance or brightness values, blue intensity values, and red intensity values.

- RGB – RGB is an image encoding method and is discussed in more detail later in this book. The pixels that make up a digital image must be represented internally in an image format and RGB is one of the available image formats used to represent these pixels. In the RGB image format each pixel contains a red, green, and blue component. The red, green, and blue components are added to get a final color. The red, green, and blue components set at the highest setting add up to white light. The red, green, and blue components set to the lowest setting represent the color black. See Figure 1-7.

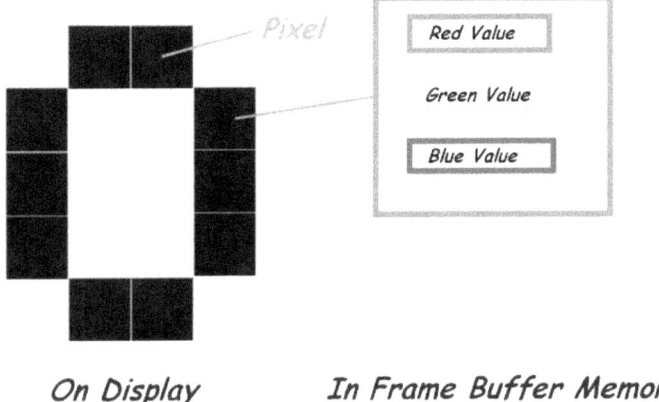

Figure 1-7. RGB image format

- Raw Bayer RGB – Raw Bayer RGB is an image format where each pixel consists of raw sensor data of either red, green, or blue depending on the color filter at that pixel location. Raw Bayer is the format the photo is initially captured in before further processing. Raw Bayer is discussed more in depth later in this book.

- Demosaicing – Demosaicing is the process by which a raw bayer RGB image can be transformed into a full color image.

- Exposure – Exposure is the amount of light per unit area and can be increased by capturing a photo over a longer period of time or decreased by capturing a photo over a shorter period of time. The end result is that the longer the exposure the brighter the captured image will be. The shorter the exposure the darker the captured image will be.

- AEC – AEC stands for Automatic Exposure Control and means that the camera will adjust the exposure setting according to certain parameters.

- AGC – AGC stands for Automatic Gain Control and controls the luminance or brightness of the photo that is taken.

- White Balancing – White Balancing is the adjustment of the colors in an image generally the colors red, green, and blue for correcting neutral colors such as gray or white so that they appear grey or white in the photo.

- AWB – AWB stands for Automatic White Balancing which means that the camera will automatically adjust the colors of the image so that neutral colors such as grey or white will appear grey or white in the captured photo.

- BLC – BLC stands for Black Level Calibration which adjusts the level of black in the image with the objective of matching true black (zero brightness) in the environment to true black in the corresponding captured image.

- ABLC – ABLC stands for Automatic Black Level Calibration that automatically adjusts the black level in the captured image according to certain parameters.

OV7670 Camera with AL422B FIFO Memory Overview

This section gives an in depth description of the ov7670 camera. First, the general capabilities of the camera are summarized. Then each functional component of the camera is described in detail. This is followed by a step by step description of how an image is captured by the camera and then transferred to the Arduino.

General Summary Of Capabilities

- Good low light operation by using NightMode

- Low operating voltage (3.3 Volts) suitable for embedded portable apps such as Arduino based projects

- Maximum frame capture rate of 30 frames per second using VGA resolution

- Compatible with Arduino though use of the camera's SCCB interface which is compatible with Arduino's I2C interface

- Supports raw Bayer RGB, RGB, YUV, and YCbCr image formats as output

- Supports VGA, QVGA, QQVGA resolutions

- Automatic image control functions including: Automatic Exposure Control, Automatic Gain Control, Automatic White Balance, Automatic Black Level Calibration.

- Supports other image processing features such as edge enhancement, denoise operations, and color correction.

- 384K (393,216) bytes frame buffer memory which is enough to hold a VGA screen capture in raw Bayer format.

Camera Functional Block Diagram

This section discusses the individual components of the Omnivision ov7670 camera. Each component is labeled with an alphabet enclosed in a circle. Each of these components is then discussed in detail. See Figure 1-8 for the full camera functional block diagram.

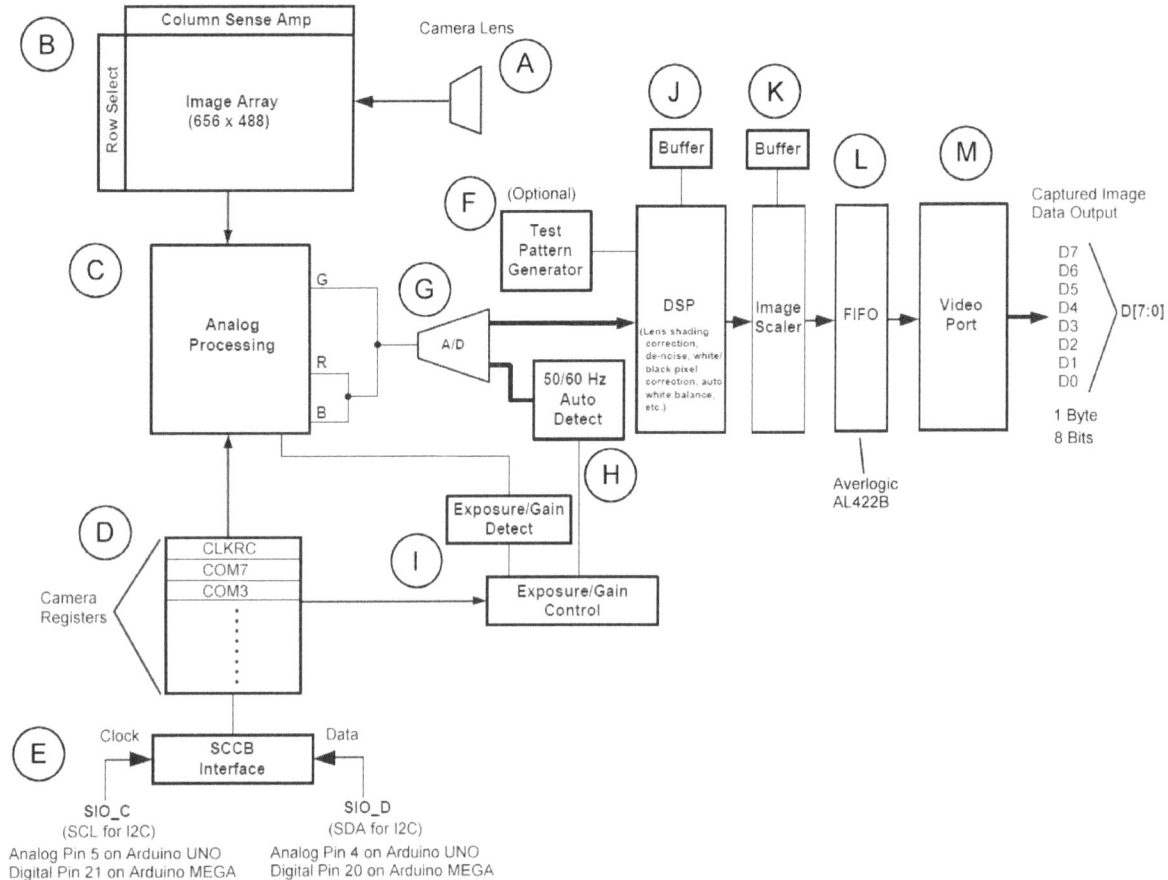

Figure 1-8. OV7670 Camera Functional Block Diagram

A. Camera Lens

The ov7670 has a lens that can be adjusted by screwing it in or out to adjust the focus of the image to be captured. Light first comes through this lens before hitting the camera's image array. See Figure 1-9.

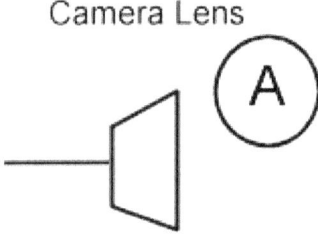

Figure 1-9. Camera Lens

B. Image Array

The camera's image array captures the incoming image and is 656 pixels wide and 488 pixels high. See Figure 1-10.

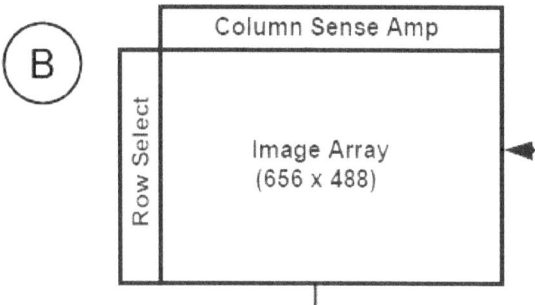

Figure 1-10. Camera's Image Array

The image array is covered with color filters arranged in a blue-green/green-red pattern. That is one row would contain color filters alternating between blue and green covering the sensor pixel cells. The next row would contain color filters alternating between green and red. See Figure 1-11.

Bayer Filter Pattern

Figure 1-11. BG/GR Bayer Filter Pattern

The way the bayer color filters work is that the red, green, and blue filters only allow in the red green, or blue component of the light reflected from the image and this intensity level is measured by the image sensors located on the pixel cells of the image array. See Figure 1-12. In case A only the red light component is measured by the pixel cell sensor. In case B only the green light component is measured by the image sensor. In case C only the blue light component is measured. Thus, each pixel in a raw bayer format image represents the intensity level of either red, green, or blue light. Each final image pixel must contain red, green, and blue information for the pixel to be correctly displayed. Therefore, the raw bayer image must go through a process called demosaicing to estimate the missing two color components needed to display the pixel correctly.

The YUV and YCbCr camera output formats use the camera's built in demosaicing algorithms to generate the final correct image. The values of each pixel in the final image are determined by the light hitting that pixel directly as well as the light hitting the surrounding pixels. The camera can also generate GRB, RGB555/RGB565 formats which are converted from YUV/YCbCr. The raw bayer images can be demosaiced using a free public domain program called FFMPEG. I discuss FFMPEG later on in this book as well as these image formats.

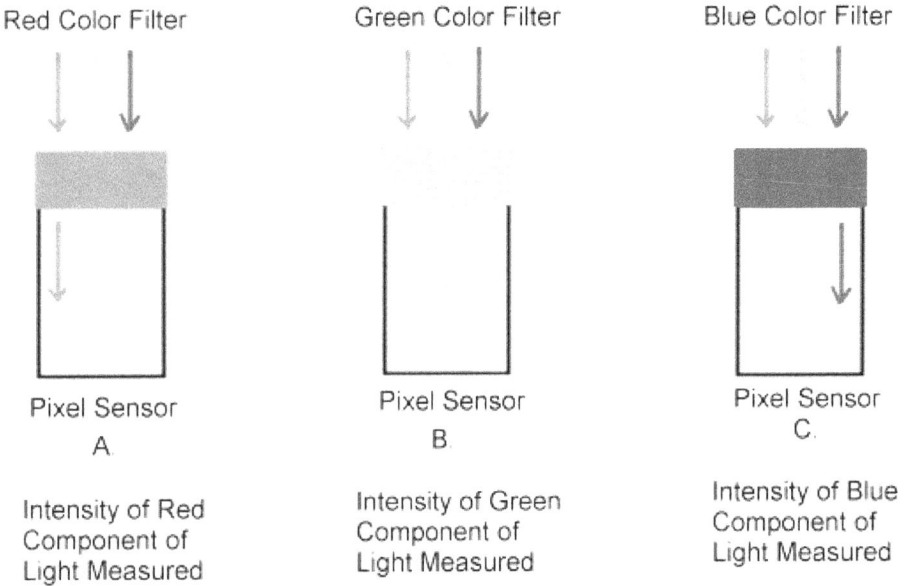

Figure 1-12. How Bayer Color Filters Work

C. Analog Processing

The ov7670's analog processing includes exposure control, gain control, and black level calibration control. Gain controls can be set to manual or automatic. The term gain refers to the luminance or brightness of the image. Setting the gain to automatic tells the camera to control the image's brightness automatically without any other control inputs supplied by the user. Black level calibration can be set to manual or automatic and adjusts the black color in the captured image as close to the actual image as possible. The exposure control can be set to manual or automatic. The (AEC) automatic exposure control methods used can be average based or histogram based. (AEC) Automatic exposure control and (AGC) automatic gain control share the same algorithms and are used together to adjust the overall luminance or brightness of the image. See Figure 1-13.

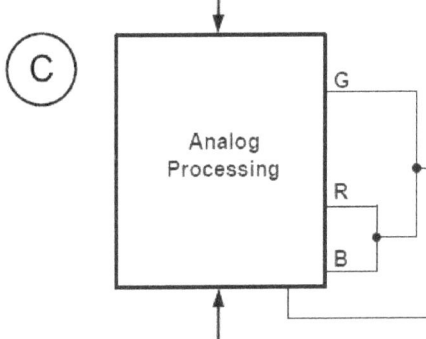

Figure 1-13. Analog Processing

The strategy in average based control of AEC and AGC involves changing the exposure and gain fast if the measured luminance is outside the control zone. Once the luminance is within the control zone the exposure and gain is changed in smaller amounts until the measured luminance is within the stable operating region. Once within the stable operating region there are no further changes to the camera's exposure and gain. See Figure 1-14.

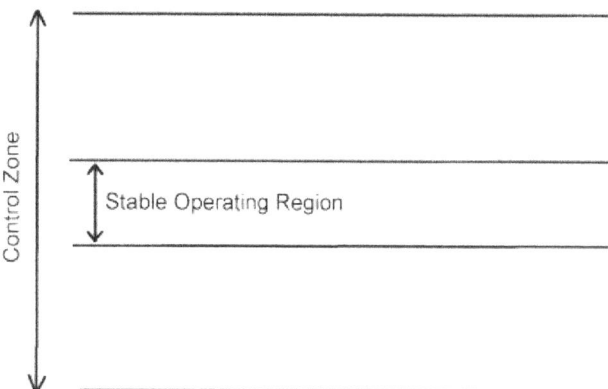

Figure 1-14. Average AEC/AGC

For the histogram method the exposure and the gain are changed until the luminance histogram reaches the desired distribution. We discuss the average based and histogram based AEC and AGC control methods more in depth later in this book.

D. Camera Registers

The registers in the ov7670 camera are memory cells that are 1 byte or 8 bits in length and hold values that are used to control the camera's functions such as resolution, image output format, exposure, gain, frame rate, etc. If you are new to digital design I discuss bytes and bits later in this book so don't worry if you are unfamiliar with these terms. You can set and read the values of these registers through the camera's SCCB interface using the Arduino. See Figure 1-15.

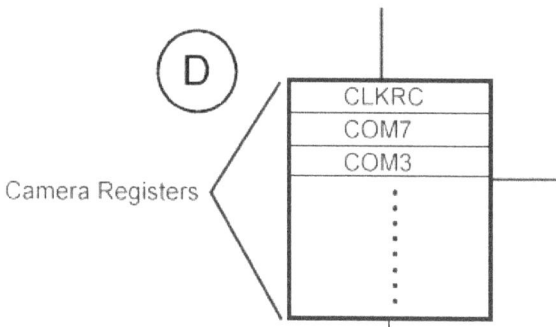

Figure 1-15. Camera Registers

E. SCCB Interface

This interface is used to read data from the camera's registers and to write data to the camera's registers. The SCCB interface on the camera is compatible with the Arduino's I2C interface and code used to activate and use the I2C interface will work with the camera's SCCB interface without any modifications. There are two pins which are the clock which is labeled the SIO_C and the data which is labeled SIO_D. The SIO_C is the same as the SCL on the I2C interface and the SIO_D is the same as the SDA on the I2C interface.

The SIO_C is connected to the Arduino UNO through analog pin 5 and is connected to the Arduino MEGA through digital pin 21. The SIO_D is connected to the Arduino UNO through analog pin 4 and connected to the Arduino MEGA through digital pin 20. See Figure 1-16.

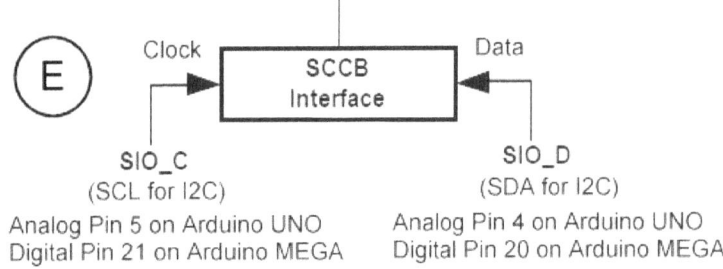

Figure 1-16. The SCCB Interface

F. Test Pattern Generator

The test pattern generator is used to display a standard set of vertical colored bars that are used to determine if the camera is working properly. Not only should a vertical group of colored bars be displayed clearly but the colors must also be in the right order. We get into more detail regarding the test pattern generator later in this book. See Figure 1-17.

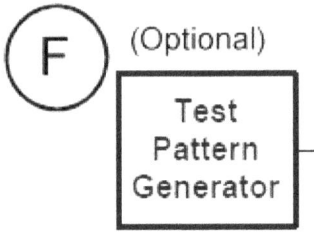

Figure 1-17. The Test Pattern Generator

G. Analog to Digital Converter

The analog to digital converter converts the raw bayer image from the image array to a digital format using a 10 bit converter. See Figure 1-18.

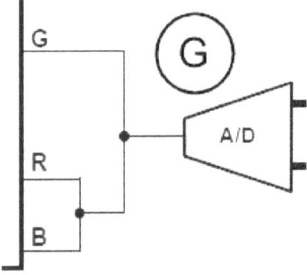

Figure 1-18. A/D Converter

H. 50/60 Hz Auto Detect

The 50/60 Hz auto detect can automatically detect the frequency of artificial light such as florescent light used in an office or home. This feature can be used with the camera's band filter features to remove any light banding that may occur in an image. See Figure 1-19.

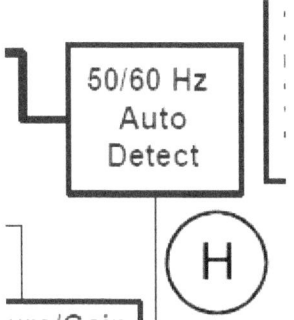

Figure 1-19. 50/60 Hz Auto Detect

I. Exposure/Gain Detection and Control

This component of the camera is responsible for detecting and controlling the exposure and gain of the image that is processed in the analog processing block. It receives control information from the camera's registers and then sets the exposure and gain accordingly. For automatic exposure and automatic gain control the camera automatically controls the exposure and gain based on the exposure and gain detected in the incoming image. See Figure 1-20.

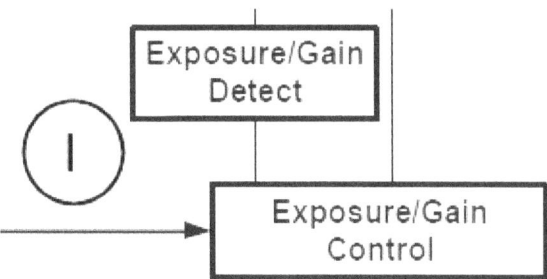

Figure 1-20. Exposure/Gain detection and control

J. Digital Signal Processor (DSP)

The digital processor or (DSP) receives digital image data from the analog to digital converter and is responsible for:

- White Balance Control

- Gamma Control

- Color Matrix

- Sharpness Control

- De-Noise
- Automatic Color Saturation Adjustment
- Defect Pixel Correction

See Figure 1-21.

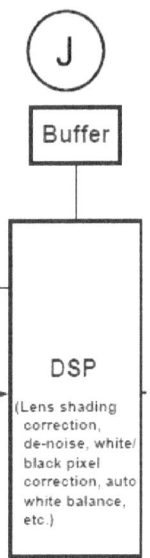

Figure 1-21. Digital Signal Processing (DSP)

White Balance Control

The white balance control for the camera allows for both manual and automatic control. The objective of white balance control is to make the white colors in the image white regardless of light color and can be set to normal (simple) mode or advanced mode.

The normal mode for automatic white balance makes the average values of the red, green, and blue colors for all the pixels in the image equal by changing the red, green, and blue gains. It assumes that the average of all the colors in the world is gray. The normal or simple mode does not depend on the characteristics of the camera lens being used to take the photo.

The advanced mode for automatic white balance uses the color temperature to adjust the red, green, and blue gains. The advanced mode depends on the characteristics of the specific lens that is being used to take the picture.

A separate pre-gain value for the red, green, and blue channels is also supported.

Gamma Control

Gamma control provides gamma correction to the image which controls its luminance or brightness. The user can set individual values that define a gamma curve that is used to lighten or darken the image.

Color Matrix

The color matrix can perform color correction and color conversion on the camera's image. The color matrix is used in conversion from raw bayer RGB to YUV/YCbCr. The matrix itself is 3 by 3 and is active in image formats YUV/YCbCr and image formats derived from YUV/YCbCr such as RGB565, RGB555, and RGB444.

Raw RGB values are converted to Cr and Cb values by multiplying the RGB value of a pixel by the ColorMatrix. The Y value is taken directly from the camera's sensor and is not affected by the ColorMatrix. See Figure 1-22.

$$\begin{bmatrix} Cr \\ Cb \end{bmatrix} = ColorMatrix \begin{bmatrix} R \\ G \\ B \end{bmatrix}$$

Figure 1-22. YCbCr color matrix use

The ColorMatrix is built from the values in camera registers MTX1, MTX2, MTX3, MTX4, MTX5, and MTX6. All that is needed is to set those registers and the camera will use the new values in converting the original picture into the final image. See Figure 1-23.

$$ColorMatrix = \begin{bmatrix} MTX1 & MTX2 & MTX3 \\ MTX4 & MTX5 & MTX6 \end{bmatrix}$$

Figure 1-23. The ColorMatrix value

The same matrix is used for conversion of RGB values to YUV. See Figure 1-24.

$$\begin{bmatrix} V \\ U \end{bmatrix} = ColorMatrix \begin{bmatrix} R \\ G \\ B \end{bmatrix}$$

Figure 1-24. YUV color matrix use

Sharpness or Edge Enhancement Control

The sharpness control can be either set to manual or automatic. The sharpness feature only works on processed bayer, YUV/YCbCr images or those that are derived from them. The raw bayer image does not contain any digital processing including sharpness or edge enhancement adjustments. If the sharpness control is set to automatic then the sharpness will vary according to a limits supplied by the user in the camera registers REG75 and REG76. In automatic mode the sharpness changes inversely with the gain. For example, the higher the gain the lower the sharpness.

De-Noise

The camera has a built in de-noise function that can be set to manual or automatic mode. In automatic mode the de-noise level is proportional to the gain. That is the greater the gain the stronger the de-noise applied to the image. The de-noise function will work on processed bayer RGB, YUV/YCbCr or any derived format such as RGB555, RGB565, and RGB444. De-noise will not work on raw bayer RGB since that format does not go through the digital signal processor.

Automatic Color Saturation Adjustment

The camera an automatically adjust color saturation based on gain. The higher the gain the weaker the color.

Defect Pixel Correction

The camera has built in pixel error correction to compensate for bad pixels on the image array.

K. Image Scaler

The image scaler reduces the size of the VGA image (if desired) that is output by the camera's digital signal processor. All other resolutions are produced by scaling down the VGA image. See Figure 1-25.

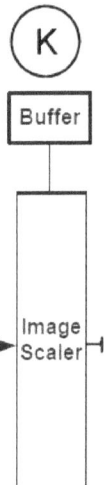

Figure 1-25. Image Scaler

L. FIFO Frame Buffer Memory

The FIFO frame buffer memory is made by AverLogic and the model is AL422B. It holds the image so that it can be read in by the Arduino. It is 384K which is enough to hold a VGA raw bayer RGB image of 1 byte per pixel with a resolution of 640 pixels wide by 480 pixels high. See Figure 1-26. An important item to be aware of is that the FIFO memory can only hold 1 byte per pixel at VGA resolution. If you attempt to write more than 1 byte per pixel to the frame buffer at VGA resolution such as trying to use the YUV image format mode with VGA then the image you get will be incorrect.

Figure 1-26. AverLogic AL422B FIFO memory

M. Video Port

The image from the FIFO memory can be output through the camera's video port. The video port is 1 byte or 8 bits in length. Thus, we will need to read the image data from the camera's video port 1 byte at a time until the entire captured image is transferred to the Arduino. See Figure 1-27.

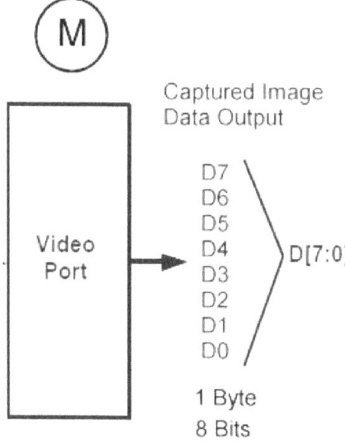

Figure 1-27. Video Port

Summary of Steps Needed for Taking a Photo

This section gives a general overview of what steps occur when an image is captured, processed and transferred to the Arduino using the Omnivision ov7670 digital camera. See Figure 1-28.

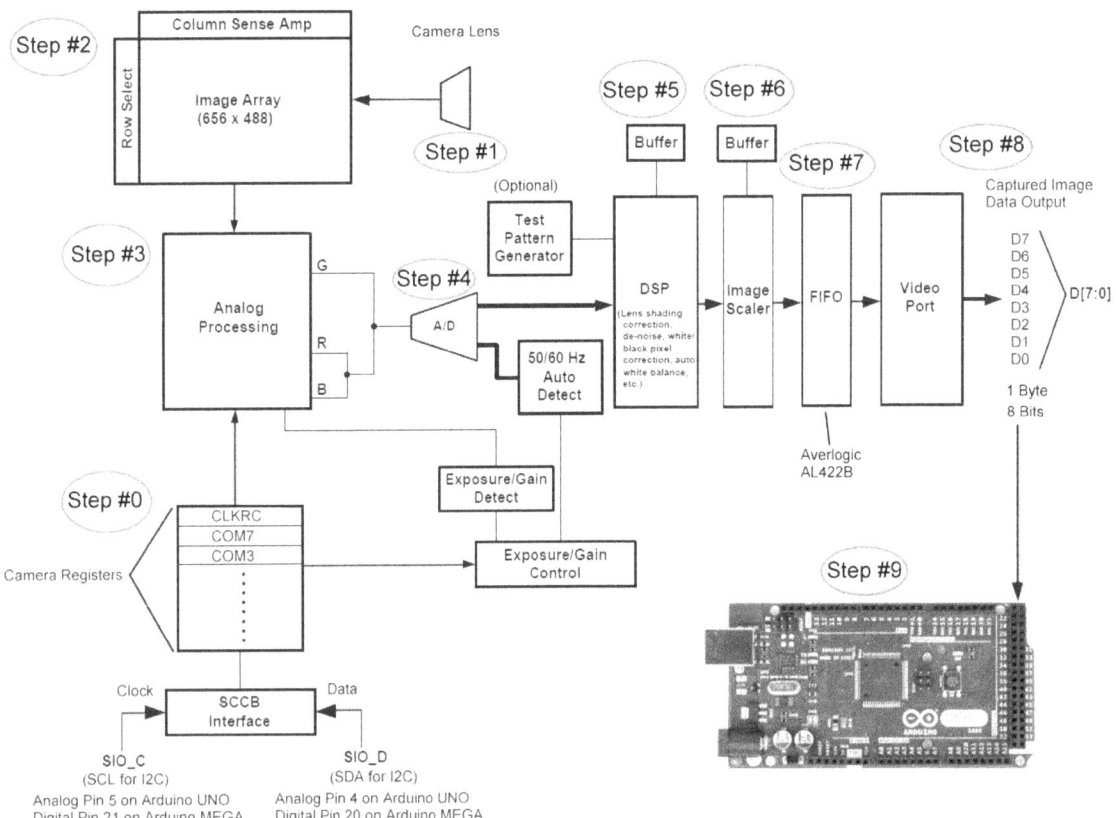

Figure 1-28. Steps in taking a photo

Step #0 – Setting the Camera's Registers

A preliminary step before capturing the image is to set the camera resolution, set the image output format, and set other image processing parameters that the user desires. This is done by the Arduino using the ov7670 camera's SCCB interface to write the required values to the camera's registers.

Step #1 – The Camera Lens

The image that is to be captured by the camera must first go through the camera's lens. It is here that the focus can be adjusted by the user by screwing the lens clockwise or counterclockwise to get a clear image.

Step #2 – The Image Array

The image array receives the incoming image after it goes through the camera's lens. Here the camera's pixel cell sensors detect the red, green, and blue components of the incoming light. These pixel cell sensors are arranged in a raw bayer image format of alternating rows of blue-green/green-red pattern.

Step #3 – Analog Processing

Next, the image goes through analog processing where the analog items like the image exposure and gain are adjusted according to the camera's register values.

Step #4 – A/D Converter

Next, the analog image is sent through the analog to digital converter that converts the image into its digital form of bytes consisting of 0's and 1's.

Step #5 – Digital Signal Processor

Then, the image is processed by the digital signal processor that handles things like white balance, edge enhancement, and de-noising.

Step #6 – Image Scaler

Next, the digitally processed image is sent to the image scaler where it is reduced in size according to the values in the camera registers that control the size of the final image that is output. Remember that all images are first captured in VGA resolution but can be scaled down using the image scaler.

Step #7 – FIFO

Then, the final scaled image is sent to the FIFO frame buffer memory which holds the image so that it can be read and output to the Arduino.

Step #8 – Video Port

Next, the video port which consists of 8 output pins representing 8 bits or 1 byte is the actual physical point where wires are attached in order to send the image data out to the device that will receive the image data.

Step #9 – Arduino

Finally, the wires from the video port on the camera are connected to pins on the Arduino designated as input pins. From there the image data is read in one byte at a time until the entire image is processed. For example, the image can be saved to a SD card or transmitted via bluetooth to an Android device to be displayed.

Summary

In this chapter I covered the Omnivision ov7670 camera. I started with a discussion of key terms and concepts relating to digital cameras that were essential in understanding the rest of the book. Then I went into a detailed discussion of the camera involving key functions and then I discussed the steps an image went through when being captured, processed and then sent from the camera to the Arduino.

Chapter 2

Introducing the Arduino

In this chapter I introduce you to the Arduino. I first give a brief explanation of what the Arduino is. I then specifically concentrate on the Arduino Mega 2560. I discuss the general features of the Arduino Mega 2560 including the capabilities and key functional components of the device. Next, I discuss the Arduino IDE (Integrated Development Environment) software that is needed to develop programs for the Arduino. I cover each key function of the Arduino IDE and then conclude with a hands on example where I give detailed step by step instructions on how to set up the Arduino for development and how to run and modify an example program using the Arduino IDE.

What is an Arduino?

The Arduino is a open source microcontroller that uses the C and C++ languages to control digital and analog outputs to devices and electronics components and to read in digital and analog inputs from other devices and electronics components for processing. For example, the Arduino can read in information from a sensor to a home security system that would detect the heat that a human being emits and sends a signal to the Arduino to indicate that a human is in front of the sensor. After receiving this information the Arduino can send commands to a camera such as the ov7670 to start taking pictures of the intruder or intruders and save these images to a SD card for later viewing. There are many different Arduino models out there. However, in order to perform the examples in this book you will need an Arduino model with enough pins to connect both the camera and the sdcard reader/writer. The official Arduino logo is shown in Figure 2-1.

Figure 2-1. Official Arduino Logo

Note: The official web site of the Arduino project is http://www.arduino.cc however in 2014 there appears to be a split between the founders of the Arduino project as to who controls the "Arduino" trademark name. Another web site called http://www.Arduino.org

was created by the company of one of the Arduino founders that split from the main group.

The Arduino Mega 2560

There are a wide range of Arduino models ranging from models that are small and can actually be worn by the user to Arduino models with many digital and analog input/output pins. For the examples in this book I recommend the Arduino Mega 2560. The Arudino Mega 2560 is an open source microcontroller that has enough digital ports to accommodate the ov7670 camera and a sd card reader/writer with enough digital and analog ports for other devices, sensors, lights, and any other gadgets that you may require for your own custom camera projects. There is an official Arduino Mega 2560 board made by a company called Arduino SRL formerly Smart Projects formed by one of the founders of the Arduino. See Figure 2-2.

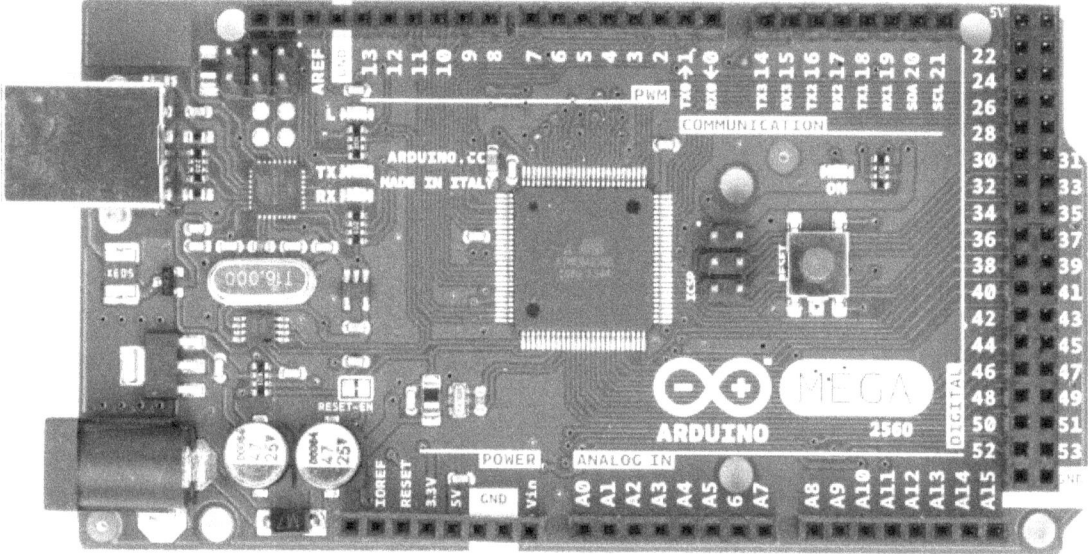

Figure 2-2. The Official Arduino Mega 2560

There are also unofficial Arduino Mega 2560 boards made by other companies. A good way to tell which board is official and which is unofficial is by the color of a component that is located near the Arduino's usb port. The component on official Arduino boards is colored a metallic gold. The component on unofficial boards has a green color. The writing on the component also differs. See Figure 2-3.

Figure 2-3 Official vs. Unofficial Arduino Boards

There are also other companies that manufacture Arduino Mega 2560 boards such as Funduino. See Figure 2-4. The name "Funduino" appears where the official Arduino "Logo" would have appeared if this was an official Arduino board. Also note the green component next to the usb port. Since the Arduino schematics are open source and other companies can legally manufacture this board there are many competing companies making this board and the boards vary in quality and price. Generally, an unofficial Arduino Mega 2560 is around $16.00 to $25.00 and an official Arduino Mega 2560 board is around $37.00 to $65.00.

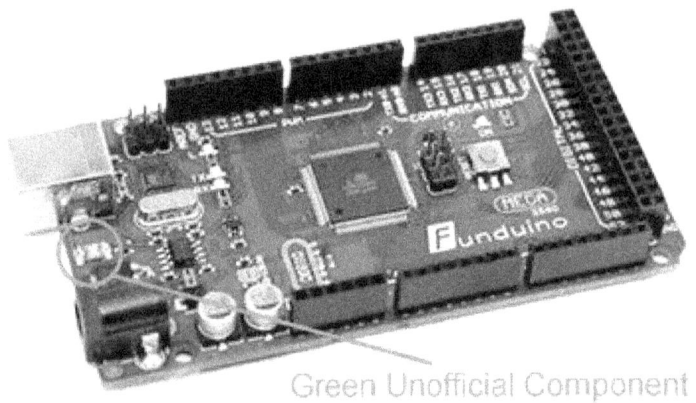

Figure 2-4. Unofficial Arduino Mega 2560 from Funduino

The Arduino Mega 2560 Specifications

Microcontroller:	ATmega2560
Operating Voltage:	5V
Input Voltage (recommended):	7-12V
Input Voltage (limits):	6-20V
Digital I/O Pins:	54 (of which 15 provide PWM output)
Analog Input Pins:	16
DC Current per I/O Pin:	40 mA
DC Current for 3.3V Pin:	50 mA
Flash Memory:	256 KB for storing code of which 8 KB used by the bootloader
SRAM:	8 KB
EEPROM:	4 KB
Clock Speed:	16 MHz

Arduino Mega 2560 Components

This section covers the functional components of the Arduino Mega 2560.

USB Connection Port

The Arduino Mega 2560 has a USB connector that is used to connect the Arduino to the main computer development system via standard USB A male to B male cable so it can be programmed and debugged. See Figure 2-5.

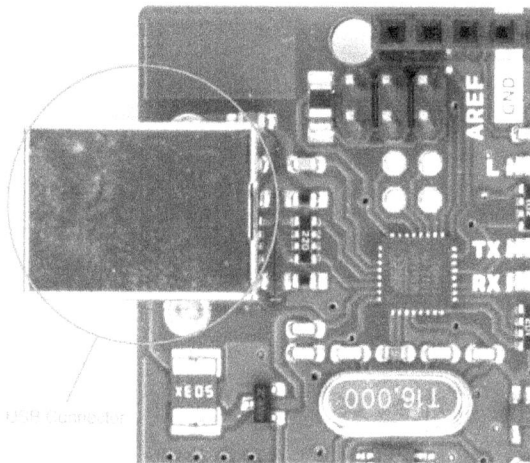

Figure 2-5. USB Connector

9V Battery Connector

The Arduino Mega 2560 has a 9 volt battery connector where you can attach a 9 volt battery to power the Arduino. See Figure 2-6.

Figure 2-6. 9 volt battery connector

Reset Button

There is a reset button on the Arduino Mega 2560 where you can press the button down to reset the board. This restarts the program contained in the Arduino's memory. See Figure 2-7.

Figure 2-7. Reset Button

Digitial Pulse Width Modulation

The Arduino Mega has many digital pins capable of simulating analog output through the process of pulse width modulation. For example, an L.E.D. light generally has only two modes which is on (full brightness) or off (no light emitted). However, with digital pulse width modulation the L.E.D. light can appear to have a brightness in between on and off. For instance, with PWM (Pulse Width Modulation) an L.E.D. can start from an off state and slowly brighten until it is at its highest brightness level and then slowly dim until back to the off state. I discuss pulse width modulation later in this book. The digital pins on the Arduino Mega 2560 that support PWM are pins 2 through pin 13. These PWM capable digital pins are circled in Figure 2-8.

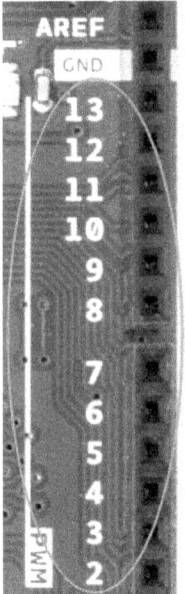

Figure 2-8. Digital pulse width modulation

Communication

The communication section of the Arduino Mega 2560 contains pins for serial communication between the Arduino and another device such as a bluetooth adapter or your personal computer. The Tx0 and the Rx0 pins are connected to the USB port and serve as communication from your Arduino to your computer through your USB cable. The Serial Monitor that can be used for sending data to the Arduino and reading data from the Arduino uses the Tx0 and Rx0 pins. Thus, you should not connect anything to these pins if you want to use the Serial Monitor to debug your Arduino programs or to receive user input. I will talk more about the Serial Monitor later in this book. In addition, the Arduino Mega 2560 has three more sets of serial communication pins that are labeled Tx1/Rx1, Tx2/Rx2, and Tx3/Rx3. See Figure 2-9.

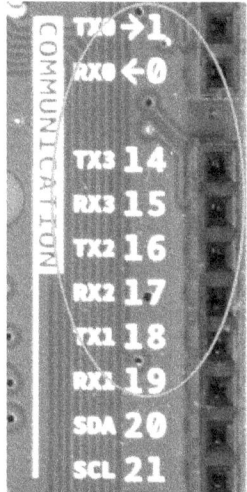

Figure 2-9. Serial Communication

The ov7670 camera will communicate with the Arduino though the I2C interface. The I2C interface consists of an SDA pin which is pin 20 and is used for data and an SCL pin which is pin 21 and is used for clocking or driving the device or devices attached to the I2C interface. The SDA and SCL pins are circled in Figure 2-10.

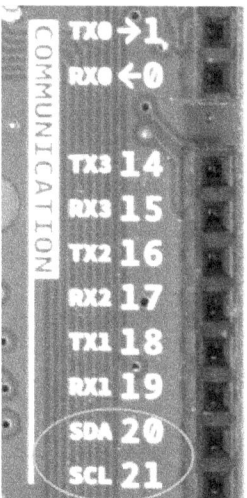

Figure 2-10. I2C Interface

Digital Output/Input

The Arduino Mega 2560 has many more digital output/input pins then the Arduino Uno which is a popular Arduino model for beginners with a limited number of digital and analog output/input pins. The camera will need many digital pins. The sd card that will be used to save the photos will also need digital pins. Pins 22 through 53 are digital pins on the Arduino Mega 2560. Pins discussed earlier that are capable of PWM (Pulse Width Modulation) are also capable of normal digital output/input. See Figure 2-11.

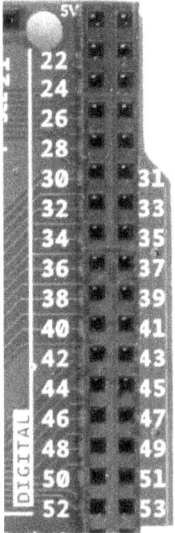

Figure 2-11. Digital Output/Input

Analog Input

The Arduino Mega 2560 has 16 analog input pins that can read in a range of values instead of just digital values of 0 or 1. The analog input pin uses a 10 bit analog to digital converter to transform voltage input in the range of 0 volts to 5 volts into a number in the range between 0 to 1023. See Figure 2-12.

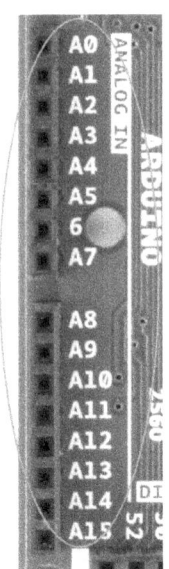

Figure 2-12. Analog Input

Power

The Arduino Mega 2560 has outputs for 3.3 volts and 5 volts. One section that provides power is located on the side of the Arduino. You can also provide your own power source by connecting the positive terminal of the power source to the Vin pin and the ground of the power source to the Arduino's ground. Make sure the voltage being supplied is within the Arduino board's voltage range. See Figure 2-13.

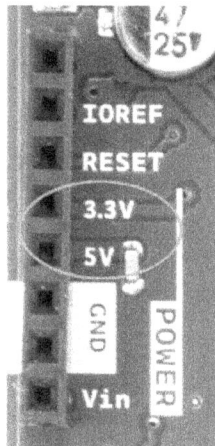

Figure 2-13. 3.3 volt and 5 volt Power outputs and Vin voltage input

Another place on the Arduino Mega 2560 that provides power is on the side of the board near the digital output pin 22 and pin 23 that provide 5 volt outputs. See Figure 2-14.

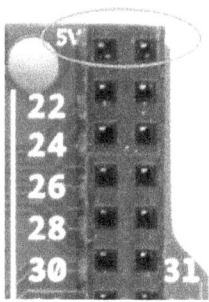

Figure 2-14. 5 volt power

Ground

The ground connections on the Arduino Mega 2560 are shown circled in Figure 2-15. We will explain more about the importance of ground connections later in this book.

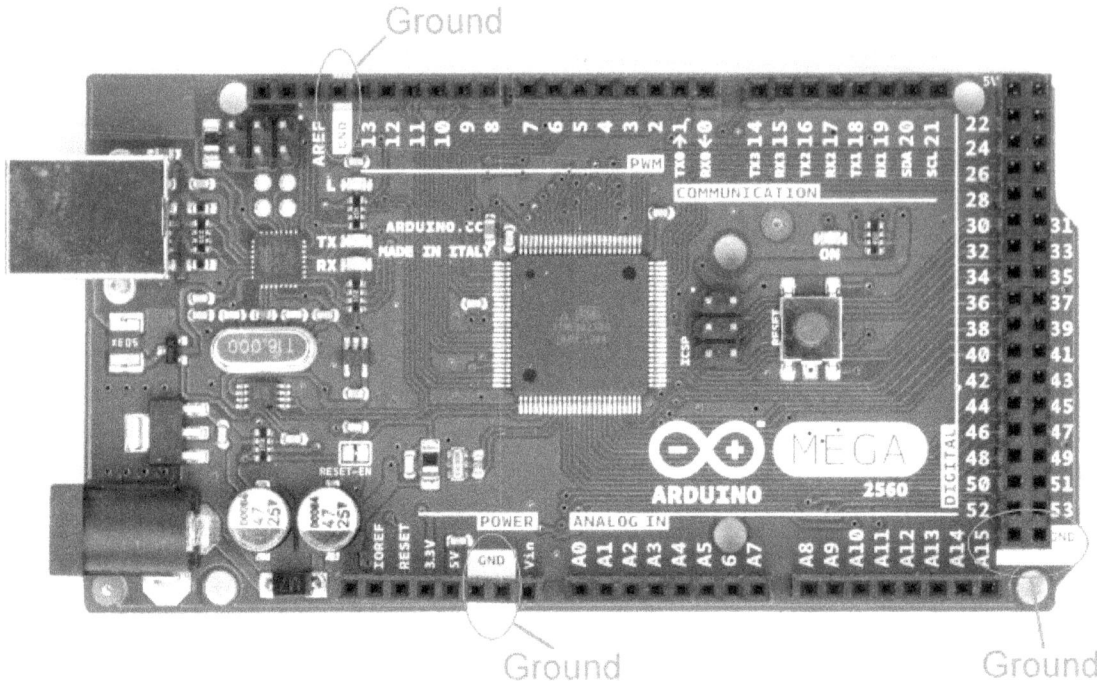

Figure 2-15. Arduino Mega 2560 Ground Connections

Arduino Development System Requirements

Developing projects for the Arduino can be done on the Windows, Mac, and Linux operating systems. The software needed to develop programs that run on the Arduino can be downloaded from the main web site at:

http://www.arduino.cc/en/Main/Software

The following is a summary of the different types of Arduino IDE distributions that are available for download. You will only need to download and install one of these files. The file you choose will depend on the operating system your computer is using.

Windows

- Windows Installer – This is a .exe file that must be run to install the Arduino Integrated Development Environment.

- Windows ZIP file for non admin install – This is a zip file that must be uncompressed in order to install the Arduino Integrated Development Environment. 7-zip is a free file compression and uncompression program available at http://www.7-zip.org

Mac

- Mac OS X 10.7 Lion or newer – This is a zip file that must be uncompressed and installed for users of the Mac operating system

Linux

- Linux 32 bits – Installation file for the Linux 32 bit operating system.

- Linux 64 bits – Installation file for the Linux 64 bit operating system.

The easiest and cheapest way to start Arduino development is probably through using the Windows version on an older operating system such as Windows XP. In fact, the examples in this book were created by using the Windows version of the Arduino IDE running on Windows XP. There are in fact many sellers on Ebay where you can purchase a used Windows XP computer for around $50-$100. So if you are starting from scratch and are looking for a inexpensive development system for the Arduino then consider buying a used Windows XP based computer. The only caution is that support for the Windows XP has ended in the United States and some other parts of the world. In China Windows XP may still be supported with software updates such as security patches.

Arduino Software IDE

As of this writing the latest Arduino IDE is version 1.6.3. This is the program that is used to develop the program code that runs on and controls the Arduino. For example, in order for you to have the Arduino control the lighting state of an L.E.D. (Light Emitting Diode) you will need to write a computer program in C/C++ using the Arduino IDE. Then, you will need to compile this program into a form that the Arduino is able to execute and then transfer the final compiled program using the Arduino IDE. From there the program automatically executes and controls the L.E.D. that is connected to the Arduino.

New versions of the IDE are compiled daily or hourly and are available for download. Older versions of the IDE are also available for downloading at:

http://www.arduino.cc/en/Main/OldSoftwareReleases

In this section we will go over the key features of the Arduino Software IDE. The IDE you are using may be slightly different then the version discussed in this section but the general functions we cover here should still be the same. We won't go in depth into every detail of the IDE since this book is meant as a quick start guide and not a reference manual. We will cover the critical features of the Arduino IDE that you will need to get started on the main camera project in this book. See Figure 2-16.

Figure 2-16. The Arduino IDE

The "verify" button checks to see if the program you have entered into the Arduino IDE is valid and without errors. These programs are called "sketches". See Figure 2-17.

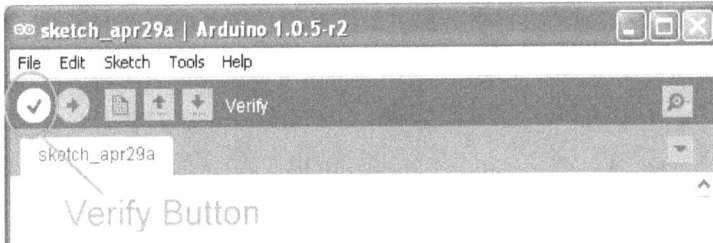

Figure 2-17. The verify button

The "upload" button first verifies that the program in the IDE is a valid C/C++ program with no errors, compiles the program into a form the Arduino can execute, and then finally transfers the program via the USB cable that is connected to your computer to your Arduino board. See Figure 2-18.

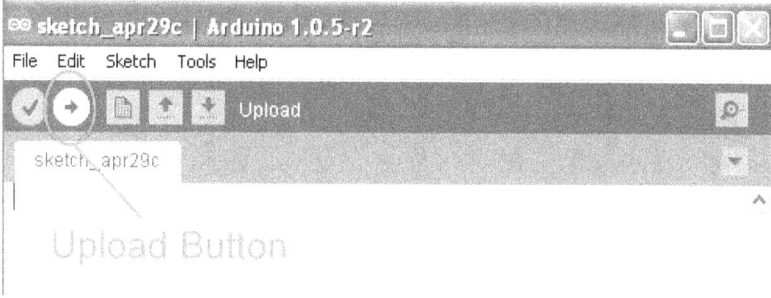

Figure 2-18. The upload button

The "New" button creates a new blank file or sketch inside the Arduino IDE where the user can create his or her own C/C++ program for verification, compilation, and transferring to the Arduino. See Figure 2-19.

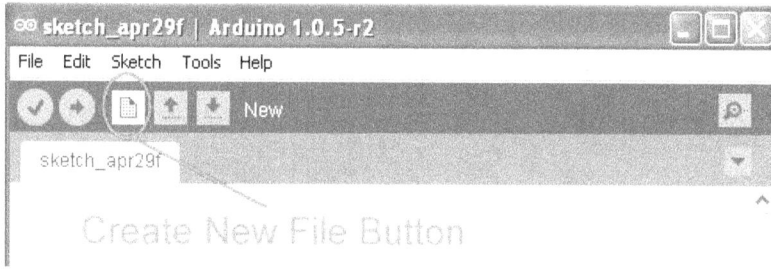

Figure 2-19. The new file button

The "open" file button is used to open and load in the Arduino C/C++ program source code from a file or load in various sample source codes from example Arduino projects that are included with the IDE. See Figure 2-20.

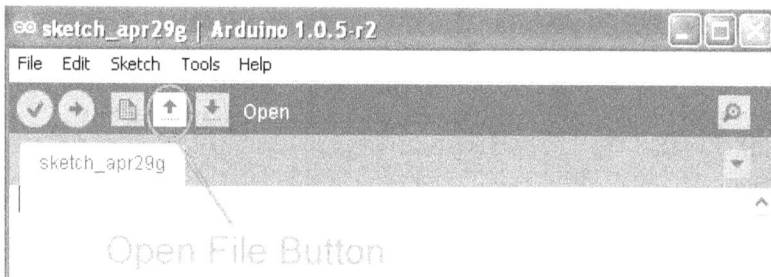

Figure 2-20. The open file button

The "save" button saves the sketch you are currently working on to disk. A file save dialog is brought up first and then you will able to save the file on your computer's hard drive. See Figure 2-21.

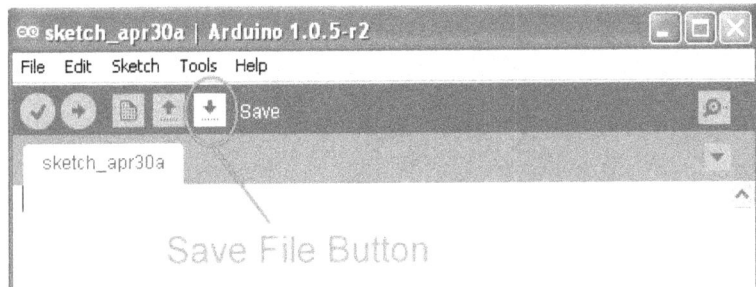

Figure 2-21. Save button

The "serial monitor" button brings up the Serial Monitor debug program where the user can examine the output of debug statements from the Arduino program. The Serial Monitor can also accept user input that can be processed by the Arduino program. We will discuss the Serial Monitor and how to use it as a debugger and input console later in this book. See Figure 2-22.

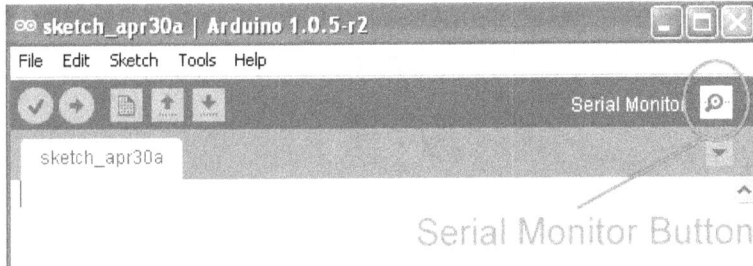

Figure 2-22. Serial Monitor

There are also other important features of the main window of the Arduino IDE. The title bar of the IDE window contains the Arduino IDE version number. In Figure 2-23 the Arduino version number is 1.0.5 r2. The sketch name is displayed in the source code tab and is "Blink" which is one of the sample sketches that come with the Arduino IDE. The source code area which is the large white area with scrollbars on the right side and bottom is where you enter your C/C++ source code that will control the behavior of the Arduino. The bottom black area in the IDE is where warning and errors are displayed from the code verification process. At the bottom left hand corner of the IDE is a number that represents the line number in the source code where the user's cursor is currently located. In the lower right hand corner of the IDE is the currently selected Arduino model and COM port that the Arduino is attached to.

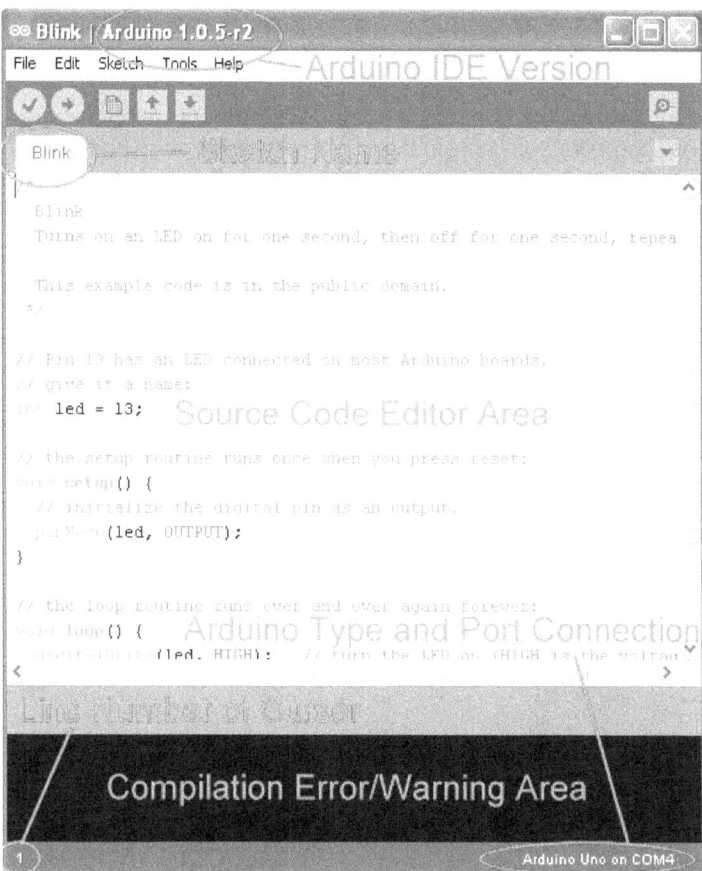

Figure 2-23. The General IDE

Hands on Example: A simple Arduino "Hello World" program with an LED

In this hands on example I show you how to set up the Arduino development system on your Windows based PC or Mac. I first discuss where you can get an Arduino board and USB cable. Then I discuss the installation of the Arduino IDE and Arduino hardware device drivers. I then discuss how to load in the "Blink" sketch example program. Next, I tell you how to verify that the program is without syntax errors, how to upload it onto the Arduino, and how to tell if the program is working. Finally, I discuss how the Blink program code works and show you how to modify it.

Get an Arduino Board and USB Cable

You can purchase an official Arduino Mega 2560 board from a distributor listed on http://www.arduino.cc/en/Main/Buy or http://www.Arduino.org. The first web site is still generally considered the main web site for Arduino. However, the second web site is run by the people actually making Arduino boards. The split among the founders of the Arduino mentioned earlier can be seen here in terms of who is designated as a distributor of an "official" Arduino board.

A second option is to buy an unofficial Arduino Mega 2560 made by a seller not listed as an "official" distributor by either arduino.cc or arduino.org. These boards are generally a lot cheaper than an "official" Arduino board. However, the quality may vary widely between manufacturers or even between production runs between the same manufacturer.

In terms of the USB cable that is used to connect the Arduino to your development computer the official Arduino board generally does not come with a cable but many unofficial boards come with short USB cables. Arduino compatible USB cables of a longer length such as 6 foot or 10 foot can be bought on Amazon.com or Ebay.com. I purchased the "Mediabridge USB 2.0 - A Male to B Male Cable (10 Feet) - High-Speed with Gold-Plated Connectors – Black" from Amazon.com for my "unofficial" Arduino Mega 2560 and its seems to work well. Make sure you get the right kind of USB cable with the right connectors on either end. The rectangular end of the USB cable is connected to your computer and the square end is connected to your Arduino. See Figure 2-24.

Figure 2-24. Arduino USB Cable

Install the Arduino IDE

The Arduino IDE has versions that can run on the Windows, Mac, and Linux operating systems. The Arduino IDE can be downloaded from:

http://www.arduino.cc/en/Main/Software

I recommend installing the Windows executable version if you have a windows based computer. Follow the directions in the pop windows.

> Note: The Arduino web site also contains links to instructions for installing the Arduino IDE for Windows, Mac, and Linux on http://www.arduino.cc/en/Guide/HomePage. The installation for Linux depends on the exact version of Linux being used.

Install the Arduino Drivers

The next step is to connect your Arduino to your computer using the USB cable. If you are using Windows it will try to automatically install your new Arduino hardware. Follow the directions in the pop up windows to install the drivers. Decline to connect to Windows Update to search for the driver. Select "Install the software automatically" as recommended. If you are using XP ignore the popup window warning about the driver not passing windows logo testing to verify its compatibility with XP.

If this does not work then instead of selecting "Install the software automatically" specify a specific driver location which is the "drivers/FTDI" directory under your main Arduino installation directory.

Loading in the Blink Arduino Sketch Example

Next, we need to load the Blink sketch example into the Arduino IDE. Click the "open" button to bring up the menu. Under "01. Basics" select the "Blink" example to load in. See Figure 2-25.

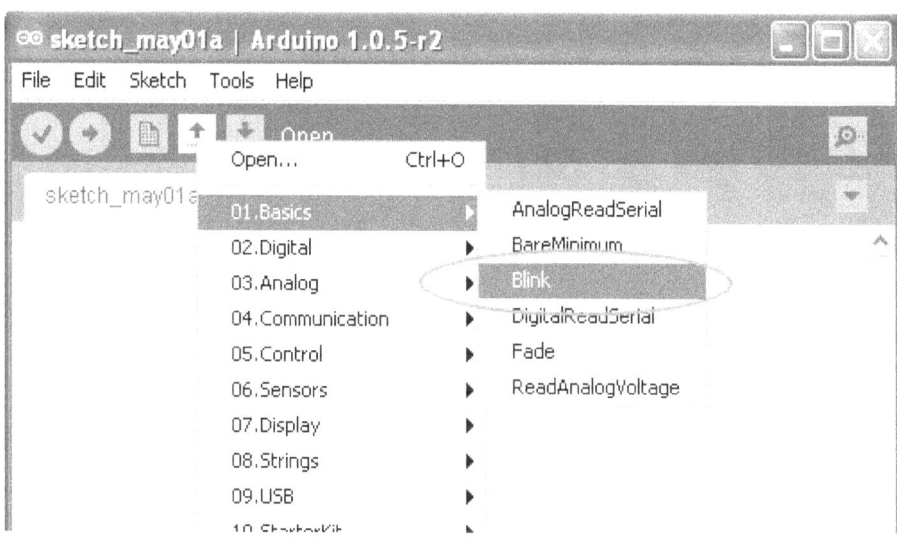

Figure 2-25. Loading in the blink example

The code that is loaded into the Arduino IDE should look like the code in Listing 2-1.

Listing 2-1. Blink Sketch

```
/*

  Blink

  Turns on an LED on for one second, then off for one second, repeatedly.



  This example code is in the public domain.

 */
```

```
// Pin 13 has an LED connected on most Arduino boards.
// give it a name:
int led = 13;


// the setup routine runs once when you press reset:
void setup() {
  // initialize the digital pin as an output.
  pinMode(led, OUTPUT);
}


// the loop routine runs over and over again forever:
void loop() {
  digitalWrite(led, HIGH);   // turn the LED on (HIGH is the voltage level)
  delay(1000);               // wait for a second
  digitalWrite(led, LOW);    // turn the LED off by making the voltage LOW
  delay(1000);               // wait for a second
}
```

Verifying the Blink Arduino Sketch Example

Click the "verify" button to verify the program is valid C/C++ code and is error free. See Figure 2-26.

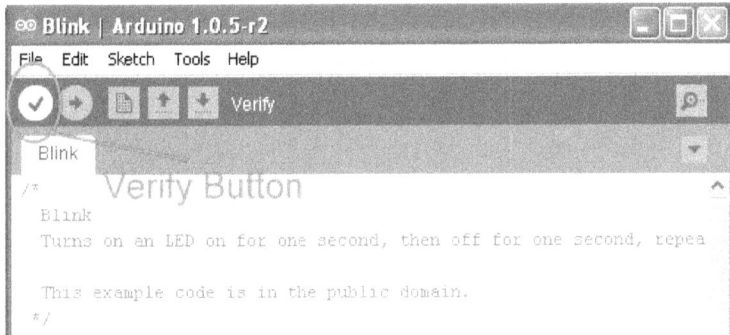

Figure 2-26. Verifying the blink sketch

Uploading the Blink Arudino Sketch Example

Before uploading the sketch to your Arduino make sure the type of Arduino under the "Tools->Board" menu item is correct. In our case the board type should be set to be "Arduino Mega 2560 or Mega ADK". See Figure 2-27.

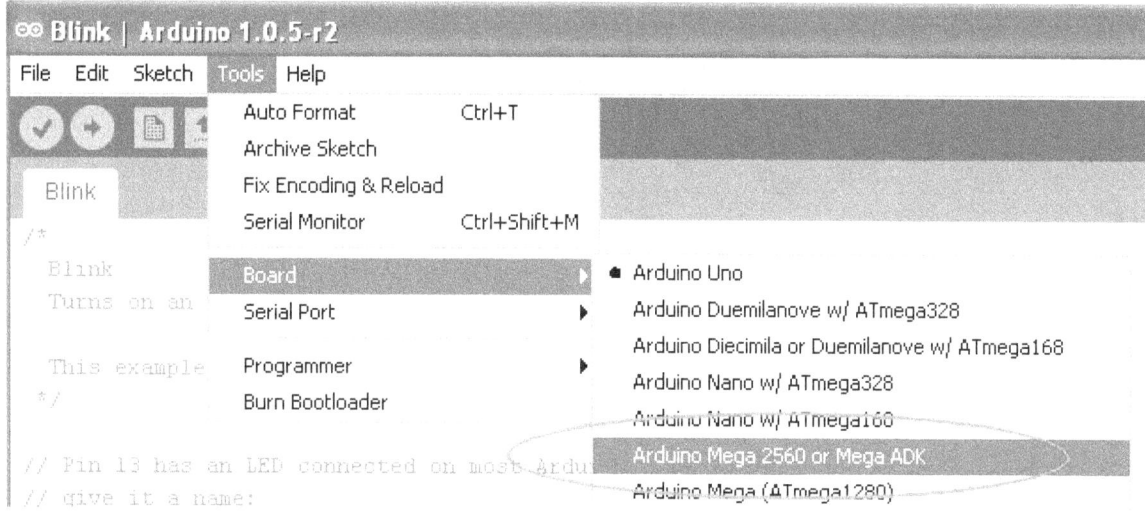

Figure 2-27. Set Arduino type to Arduino Mega 2560

Next, make sure the serial port is set correctly to the one that is being used by your Arduino. Generally Com1, and Com2 are reserved and the serial port that the Arduino will be connected to is Com3 or higher. See Figure 2-28.

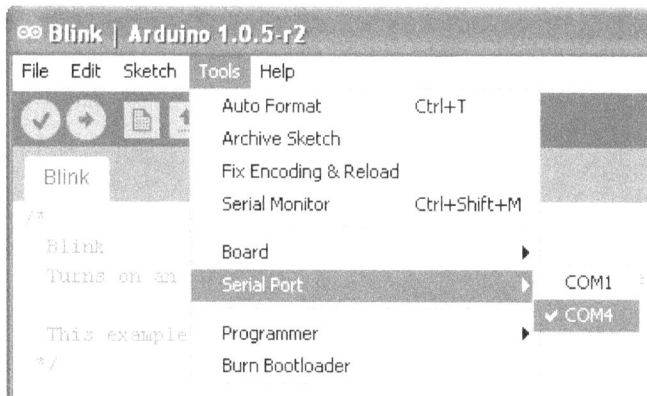

Figure 2-28. Set the com port

If you are using a Mac then the Serial Port selection should be something like "/dev/tty.usbmodem" instead of a COMXX value.

Next, with the Arduino connected press the "upload" button to verify, compile, and then transfer the Blink example program to the Arduino. After the program has finished uploading you should see a message that the upload has been completed in the warnings/error window at the bottom of the IDE inside the black window. See Figure 2-29.

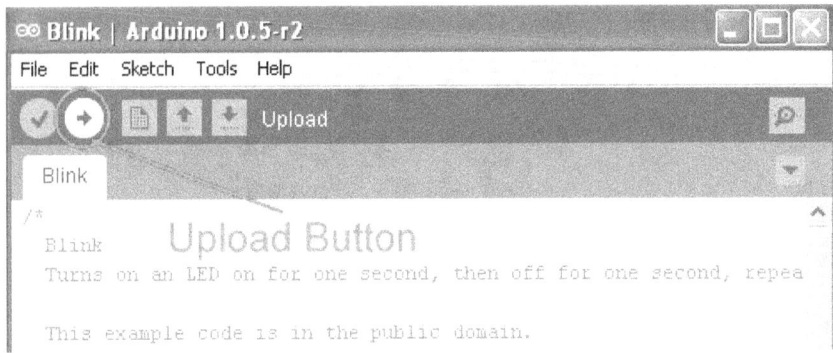

Figure 2-29. Upload to Arduino

Note: The Upload button does the job of the "verify" button but also uploads the final compiled program to the Arduino.

The Final Result

The final result will be a blinking light on the Arduino board near digital pin 13. By design the Arduino board has a built in L.E.D. connected to pin 13. So this example did not require you to connect an actual separate L.E.D. to the Arduino board. See Figure 2-30.

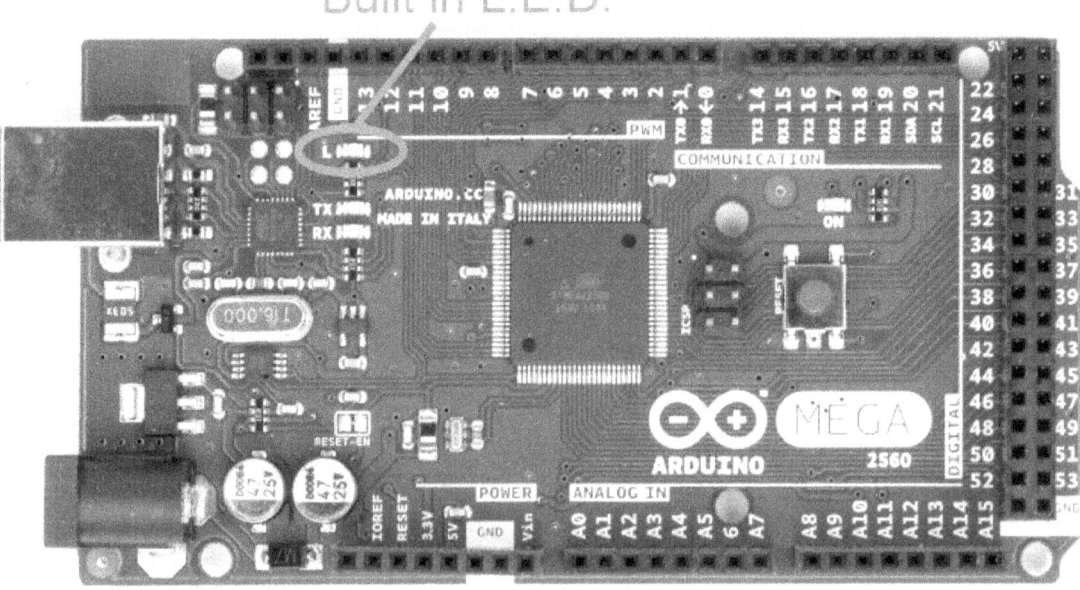

Figure 2-30. Built in L.E.D.

Playing Around with the Code

The default of the program is to turn on the L.E.D. for one second and then turn off the L.E.D. for one second. The code that controls the timing is located in the loop() function. The digitalWrite() function sets the variable led which is pin 13 to either on which is HIGH or off which is LOW. The delay() function suspends the execution of the program for 1000 milliseconds or 1 second so that the L.E.D. is set on for 1 second and off for 1 second. See Listing 2-2.

Listing 2-2. loop() Function

```
void loop() {

  digitalWrite(led, HIGH);   // turn the LED on (HIGH is the voltage level)

  delay(1000);               // wait for a second

  digitalWrite(led, LOW);    // turn the LED off by making the voltage LOW

  delay(1000);               // wait for a second

}
```

Play around with the values in the delay() functions lengthening or shortening the time the L.E.D. stays on and/or lengthening or shortening the time the L.E.D. stays off. For example, to have the L.E.D. briefly flash shorten the first delay value to 100. This would shorten the time that the L.E.D. stays on. Upload the new sketch to the Arduino by pressing the "upload" button. After it has finished uploading you see a message indicating that in the black warnings/error message window at the bottom of the IDE. Look at the L.E.D. on the Arduino. The timing of the L.E.D. on/off pattern should have changed.

Summary

In this chapter I introduced the Arduino to the reader. I concentrated my coverage on the Arduino Mega 2560. I discussed the Mega 2560's basic features and then covered key functional components. Next, information about the software needed to develop programs for the Arduino called the Arduino IDE (Integrated Development Environment) was presented including the different versions of the IDE available for different platforms. Information about key features of the IDE was discussed. Finally, a hands on example was presented where I took the reader through a step by step guide to setting up the Arduino with a development computer system. In this example I also discussed loading in an example program, uploading this example program to the Arduino, and then modifying the program to see how this changes the output.

Chapter 3

Arduino Programming Language Basics

In this chapter I go over the basics of the Arduino Language. I cover the basic elements of the Arduino language that you will need in order to create programs that control the Arduino board. Various key elements such as data types, constants, control loops, etc. are covered.

C/C++ Language for Arduino Overview

The Arduino uses C and C++ in its programs called "sketches". This section briefly summaries key language elements. This is not meant as a reference guide and ideally you should have some experience with a programming language similar to C and/or C++. Many examples, from this section come directly from the image capture program that is covered later in this book.

Comments

- // - This is a single line comment that is used by the programmer to document the code. These comments are not executed by the Arduino.

- /* */ - These enclose a multi line comment that is used by the programmer to document the code. These comments are not executed by the Arduino.

Data Types

- void – This return type is only applicable to functions and indicates that the function will not return any value to the function caller. For example, the setup() function that is part of the standard Arduino code framework has a return type of void.

    ```
    void setup()

    {

        // Initialize the Arduino, camera, and SD card here

    }
    ```

- boolean – A boolean variable can hold either the value of true or false and is 1 byte in length. For example, in the following code the variable result is declared of type boolean and is initialized to false.

  boolean result = false;

- char – The char variable type can store character values and is 1 byte in length. The following code declares that tempchar is of type char and is an array with 50 elements.

  char tempchar[50];

- unsigned char – The unsigned char data type holds 1 byte of information in the range of 0 through 255.

- byte – The byte data type is the same as the unsigned char data type. The following code declares a variable called data of type byte that is initialized to 0.

  byte data = 0;

- int – The int data type holds a 2 byte number in the range of -32,768 to 32,767. The following code declares a variable of type integer that holds the error status after writing to the camera.

  int result = OV7670WriteReg(COM7, COM7_VALUE_RESET);

- unsigned int – This data type is 2 bytes in length and holds a value from 0 to 65,535.

- word - This data type is the same as the unsigned int type.

- long – This data type is 4 bytes in length and holds a value from -2,147,483,648 to 2,147,483,647.

- unsigned long – This data type is 4 bytes in length and holds a value between 0 to 4,294,967,295.

- Float – This is a floating point number that is 4 bytes in length and holds a value between -3.4028235E+38 to 3.4028235E+38.

- double - On the current Arduino implementation the double is the same as float with no gain in precision.

- String – This is a class object that allows the user to easily manipulate groups of characters. In the following code a new variable called Command of type String is declared and initialized to the QQVGA resolution.

  String Command = "QQVGA";

- array – An array is a continuous collection of data that can be accessed by an index number. Arrays are 0 based so that the first element in the array has an index of 0. Common types of arrays are character arrays, and integer arrays. The following code declares the variable Entries as an array of type String that contains 10 elements. The function ProcessRawCommandElement() is then called with element number 2 in the Entries array which is the third element in the array. Remember 0 is the first element in the array.

    String Entries[10];

    boolean success = ProcessRawCommandElement(Entries[2]);

Constants

- INPUT – This is an Arduino pin configuration that sets the pin as an input pin that allows you to easily read in the voltage value at that pin with respect to ground on the Arduino. For example, the following code sets the pin VSYNC on the Arduino as an INPUT pin which allows you to read in the voltage value of the pin to determine if a new frame is being captured by the camera. The function pinMode() is an Arduino function included in the built in library.

    pinMode(VSYNC, INPUT);

- OUTPUT – This is an Arduino pin configuration that sets the pin as an output pin that allows you to drive other electronics components such as an L.E.D. or to provide digital input to other devices in terms of HIGH or LOW voltages. In the following code the pin WEN is set to OUTPUT using the built in pinMode() function.

    pinMode(WEN , OUTPUT);

- HIGH (pin declared as INPUT) - If a pin on the Arduino is declared as an INPUT then when the digitalRead() function is called to read the value at that pin then a HIGH value would indicate a value of 3 volts or more at that pin.

- HIGH (pin declared as OUTPUT) – If a pin on the Arduino is declared as an OUTPUT then when the pin is set to HIGH with the digitalWrite() function then the pin's value is 5 volts.

- LOW (pin declared as INPUT)- If a pin on the Arduino is declared as an INPUT then when the digitalRead() function is called to read the value at that pin then a LOW value would indicate a value of 2 volts or less.

- LOW (pin declared as OUTPUT) - If a pin on the Arduino is declared as an OUTPUT then when the digitalWrite() function is called to set the pin to LOW the voltage value at that pin would be set to 0 volts.

- true – true is defined as any non zero number such as 1, -1, 200, 5, etc.

- false – false is defined as 0.

The Define Statement

The define statement assigns a name to a constant value. During the compilation process the compiler will replace the constant name with the constant value.

#define constantName value

In the image capture program that is discussed later in this book I use the #define statement many times to assign a number to a pin label such as the Arduino pin connected to the write reset pointer pin on the camera which is pin 22.

#define WRST 22

Define statements are also used to assign numerical values to camera registers such as CLKRC.

#define CLKRC   0x11

The values of these camera registers are also assigned values using the #define statement. The following statement assigns 0x01 as the value to set the CLKRC register to put the camera into VGA resolution.

#define CLKRC_VALUE_VGA   0x01

The Include Statement

The #include statement brings in code from outside files and "includes" them into your Arduino sketch. Generally a header or .h file is included which allows access to the functions and classes inside that file.

For example, in our image capture program we include the Wire.h file which let's us use the Wire library. The Wire library has functions to initialize, to read data from and to write data to a device connected to the I2C interface. We need the Wire library to use the ov7670 camera that uses the I2C bus.

#include <Wire.h>

Also, we need to include the SD card reader library in order to use the SD card reader/writer.

#include <SD.h>

The Semicolon

Each statement in C/C++ needs to end in a semicolon. For example, when declaring and initializing a variable you will need a semicolon.

const int chipSelect = 48;

When you use a library that you included with the #include statement you will need a semicolon at the end when you call a function.

Wire.begin();

Curly Braces

The curly braces such as { and } specify blocks of code and must match in pairs. That is, for every opening brace { there must be a closing brace } to match.

A function requires curly braces to denote the beginning and end of the function.

void Function1()

{

  // Body of Function

}

Program loops such as the for statement may also need the curly braces

for (int I = 0; I < 9; I++)

{

  // Body of loop

}

It is also good practice to use braces in control structures such as the if statement.

if (I < 0)

{

  // Body of If statement

}

Arithmetic Operators

- = - The equals sign is the assignment operator used to set a variable to a value. For example, the following sets the value of the variable Data to the result from the function CreatePhotoInfo():

    String Data = CreatePhotoInfo();

- \+ - The plus sign performs addition. For example, the following adds the Strings Command, PhotoTakenCount and Ext together to get a final string called Filename.

    String Filename = Command + PhotoTakenCount + Ext;

- - - The minus sign performs subtraction. For example, the following calculates the time it takes to capture a photo using the camera by measuring the difference between the starting time before the image is captured and the ending time just after the image is captured.

    ElapsedTime = TimeForCaptureEnd - TimeForCaptureStart;

- * - The asterisk sign performs multiplication. For example, the total bytes of an image is calculated by multiplying the width of the image by the height of the image by the bytes per pixel in the image.

    int TotalBytes = ImageWidth * ImageHeight * BytesPerPixel;

- / - The back slash sign performs division. For example, the speed in miles per hour of an object is calculated by dividing the number of miles the object has traveled by the number of hours that it took to travel that distance.

    float Speed = NumberMiles / NumberHours;

- % - The percent sign is the modulo operator that returns the remainder from a division between two integers. For example,

    int remainder = dividend % divisor;

Comparison Operators

- == - The double equal is a comparison operator to test if the argument on the left side of the double equal sign is equal to the argument on the right side. If the arguments are equal then it evaluates to true. Otherwise it evaluates to false. For example, if the automatic black level correction parameter is set to off then the code block is executed.

    ```
    if (ABLCParam == "AblcOFF")
    {
            // Execute code
    }
    ```

- != - The exclamation point followed by an equal sign is the not equal to operator that evaluates to true if the argument on the left is not equal to the argument on the right side. Otherwise, it evaluates to false. For example, in the following code if the current camera resolution is not set to VGA then the code block is executed.

    ```
    if (Resolution != VGA)
    {
            // If current resolution is not VGA then set camera for VGA
    }
    ```

- < - The less than operator evaluates to true if the argument on the left is less than the argument on the right. For example, in the code that follows the for loop will execute the code block while the height is less than the height of the photo. When the height counter becomes equal or greater than the photo's height then the loop exits.

  for (int height = 0; height < PHOTO_HEIGHT; height++)

  {

      // Process every row of the photo

  }

- > - The greater than operator evaluates to true if the argument on the left side is greater than the argument on the right side. For example, in the following code if there are available characters to read in from the Serial Monitor then execute the code block. That is the number of available characters to read in must be greater than 0.

  if (Serial.available() > 0)

  {

      // Process available characters from Serial port

  }

- <= The less than sign followed by the equal sign returns true if the argument on the left hand side is less than or equal to the argument on the right hand side. It returns false otherwise.

- >= The greater than sign followed by the equal sign returns true if the argument on the left is greater than or equal to the argument on the right. It returns false otherwise.

Boolean Operators

- && - This is the "and" boolean operator that only returns true if both the arguments on the left and right side evaluate to true. It returns false otherwise. For example, in the following code only if the user selects both denoise and edge enhancement will the code block be executed. Otherwise it will not be executed.

  if ((DenoiseParam == "DenoiseYes") &&

    (EdgeParam == "EdgeYes"))

  {

      // Set camera for both denoise and edge enhancement of images.

  }

- || - This is the "or" operator and returns true if either the left argument or the right argument evaluates to true. Otherwise, it returns false. For example, in the following code if either the camera's command is set to QQVGA or QVGA then the code block is executed. Otherwise it is not executed.

    if ((Command == "QQVGA") || (Command == "QVGA"))

    {

    　　// Code

    }

- ! – The not operator returns the opposite boolean value. The not value of true is false which is 0 and the not value of false is true which is non zero. In the following code a file is opened on the SD card and a pointer to the file is returned. If the pointer to the file is NULL which has a 0 value then not NULL would be 1 which is true. The if statement is executed when the argument is evaluated to true which means that the file pointer is NULL. This means that the open operation has failed and an error message needs to be displayed.

    // Open File

    InfoFile = SD.open(Filename.c_str(), FILE_WRITE);

    // Test if file actually open

    if (!InfoFile)

    {

    　　Serial.println(F("\nCritical ERROR ... Can not open Photo Info File for output ... "));

    　　return;

    }

Bitwise Operators

- & - This is the bitwise "and" operator between two numbers where each bit of each number has the "and" operation performed on it to produce the result in the final number. The resulting bit is 1 only if both bits in each number is 1. Otherwise the resulting bit is 0.

- | - This is the bitwise "or" operator between two numbers where each bit of each number has the "or" operation performed on it to produce the result in the final number. The resulting bit is 1 if the bit in either number is 1. Otherwise the resulting bit is 0.

- ^ - This is the bitwise "xor" operator between two numbers where each bit of each number has the "xor" operation performed on it to produce the result in the final number. The resulting bit is 1 if the bits in each number are different and 0 otherwise.

- ~ - This is the bitwise "not" operator where each bit in the number following the "not" symbol is inverted. The resulting bit is 1 if the initial bit was 0 and the bit is 0 if the initial bit was 1.

- << - This is the bitshift left operator where each bit in the left operand is shifted to the left by the number of positions indicated by the right operand. For example, in the code below a 1 is shifted left PinPosition times and the final value is assigned to the variable ByteValue.

    ByteValue = 1 << PinPosition;

- >> - This is the bitshift right operator where each bit in the left operand is shifted to the right by the number of positions indicated by the right operand. For example, in the code below bits in the number 255 are shifted to the right PinPosition times and the final value is assigned to the variable ByteValue.

    ByteValue = 255 >> PinPosition;

Compound Operators

- ++ - This is the increment operator. The exact behavior of this operator also depends if it is placed before the variable being incremented or after the variable being incremented. In the following code the variable PhotoTakenCount is incremented by 1.

    PhotoTakenCount++;

If the increment operator is placed after the variable being incremented then the variable is used first in the expression it is in before being incremented. For example, in the code below the height variable is used first in the for loop expression before it is incremented. So the first iteration of the for loop below would use height = 0. The height variable would be incremented after being used in the expression.

    for (int height = 0; height < PHOTO_HEIGHT; height++)
    {
        // Process row of image
    }

If the increment operator is placed before variable being incremented then the variable is incremented first then it is used in the expression that it is in. For example, in the code below the height variable is incremented first before it is used in the for loop. This means that in the first iteration of the loop the height variable is 1 instead of 0.

    for (int height = 0; height < PHOTO_HEIGHT; ++height)
    {
        // Process row of image
    }

- `--` - The decrement operator decrements a variable by 1 and its exact behavior depends on the placement of the operator either before or after the variable being decremented. If the operator is placed before the variable then the variable is decremented before being used in an expression. If the operator is placed after the variable then the variable is used in an expression before it is decremented. This follows the same pattern as the increment operator discussed previously.

- `+=` - The compound addition operator adds the right operand to the left operand. This is actually a short hand version of

  operand1 = operand1 + operand2;

  Which is the same as the version that uses the compound addition operator.

  operand1 += operand2;

- `-=` - The compound subtraction operator subtracts the operand on the right from the operand on the left. For example, the code for a compound subtraction would be:

  operand1 -= operand2;

  This is the same as the following:

  operand1 = operand1 - operand2;

- `*=` - The compound multiplication operator multiplies the operand on the right by the operand on the left. The code for this is as follows.

  operand1 *= operand2;

  This is also equivalent to:

  operand1 = operand1 * operand2;

- `/=` - The compound division operator divides the operand on the left by the operand on the right. For example,

  operand1 /= operand2;

  This is equivalent to:

  operand1 = operand1 / operand2;

- `&=` - The compound bitwise and operator is equivalent to:

  x = x & y;

- `!=` - The compound bitwise or operator is equivalent to:

  x = x | y;

Pointer Access Operators

- * - The Dereference Operator allows you to access the contents that a pointer points to. For example, the code that follows declares a variable called pdata as a pointer to a byte and creates storage for the data using the new command. The pointer variable called pdata is then dereferenced to allow the actual data that the pointer points to be set to 1.

    byte *pdata = new byte;

    *pdata = 1;

- & - The Address Operator creates a pointer to a variable. For example, the following code declares a variable called data of type byte and assigns the value of 1 to it. A function called FunctionPointer() is defined that accepts as a parameter a pointer to a byte. In order to use this function with the variable data we need to call that function with a pointer to the variable data.

    byte data = 1;

    void FunctionPointer(byte *data)

    {

      // body of function

    }

    FunctionPointer(&data);

Variable Scope

- Global variables – In the Arduino programming environment global variables are variables that are declared outside any function and before they are used. For example in the following code below taken from the image capture program I wrote for this book the following are global variables that are declared at the beginning of the program.

    // VGA Default

    int PHOTO_WIDTH = 640;

    int PHOTO_HEIGHT = 480;

    int PHOTO_BYTES_PER_PIXEL = 2;

- Local variables – Local variables are declared inside functions or code blocks and are only valid inside that function or code block. For example, in the following function the variable localnumber is only visible inside the Function1() function.

    void Function1()

    {

```
        int localnumber = 0;

    }
```

Conversion

- char(x) – This function converts a value x into a char data type and then returns it.

- byte(x) – This function converts a value x into a byte data type and then returns it.

- int(x) – This function converts a value x into a integer data type and then returns it.

- word(x) – This function converts a value x into a word data type and then returns it.

- word(highbyte,lowbyte) – This function combines two bytes, the high order byte and the low order byte into a single word and then returns it.

- long(x) – This function converts a value x into a long and then returns it.

- float(x) – This function converts a value x into a float and then returns it.

Control Structures

- if (comparison operator) – The if statement is a control statement that tests if the result of the comparison operator or argument is true. If it is true then execute the code block. For example, in the following code the if statement tests to see if the YUV matrix parameter is set to activate the YUV color matrix. If it is set then the SetCameraColorMatrixYUV() function is called.

    ```
    // Set Color Matrix for YUV

    if (YUVMatrixParam == "YUVMatrixOn")

    {

        SetCameraColorMatrixYUV();

    }
    ```

- if (comparison operator) else – The if else control statement is similar to the if statement except with the addition of the else section which is executed if the previous if statement evaluates to false and is not executed. For example, in the following code if the frames per second parameter is set for 30 frames per second then the SetupCameraFor30FPS() function is called. Otherwise, if the frames per second parameter is set to night mode then the SetupCameraNightMode() function is called.

    ```
    // Set FPS for Camera
    ```

```
if (FPSParam == "ThirtyFPS")
{
    SetupCameraFor30FPS();
}
else
if (FPSParam == "NightMode")
{
    SetupCameraNightMode();
}
```

- for (initialization; condition; increment) – The for statement is used to execute a code block usually initializing a counter then performing actions on a group of objects indexed by that incremented value. See the code example below.

```
for (int i = 0 ; i < NumberElements; i++)
{
    // Process Element i  Here
}
```

- while (expression) – The while statement executes a code block repeatedly until the expression evaluates to false. In the following code the while code block is executed as long as data is available for reading from the file.

```
// read from the file until there's nothing else in it:
while (TempFile.available())
{
    Serial.write(TempFile.read());
}
```

- break – A break statement is used to exit from a loop such as a while or for loop. In the following code the while loop causes the code block to be executed forever. If there is data available from the Serial Monitor then it is processed and then the while loop is exited.

```
while (1)
{
        if (Serial.available() > 0)
        {
```

```
            // Process the data

        break;

    }
}
```

- return (value) – The return statement exits a function. It also may return a value to the calling function.

    return;

    return false;

Summary

In this chapter I covered the basics of the Arduino programming language. I covered a broad range of basic topics such as data types, constants, built in functions, and control loops. In addition, this chapter was not meant to be a reference manual but a quick start guide to the basics of the Arduino programming language. Please refer to the official Arduino language reference for more information.

Chapter 4

Digital Design Review

In this chapter I cover basic digital design. I start by discussing how data is stored in the camera. I cover the different types of number systems such as decimal (base 10), binary (base 2), and hexadecimal (base 16). I then discuss how to convert numbers between these systems. Next, I give a hands on example of how to set the camera's color bar test feature by setting a register. Boolean variables, boolean logic, and boolean operators are discussed. Next, the clock pulse that drives digital circuits is covered. The schematic symbols for digital operators such as AND, OR, and NOT are then discussed. Finally, a design overview of the entire camera system at a chip level is given. The main camera chip, and the FIFO frame buffer memory chip are discussed in detail.

How Data is Stored in the ov7670 Camera

Data is organized in a device such as a computer or camera in bits and bytes. A bit a single memory cell that can hold a value that represents a 0 or 1. Eight continuous bits together form 1 byte of memory. The bits in a byte are numbered bit 0 to bit 7 with the bit at the rightmost place designated as the 0 bit and the bit at the leftmost place designated the 7th bit. See Figure 4-1.

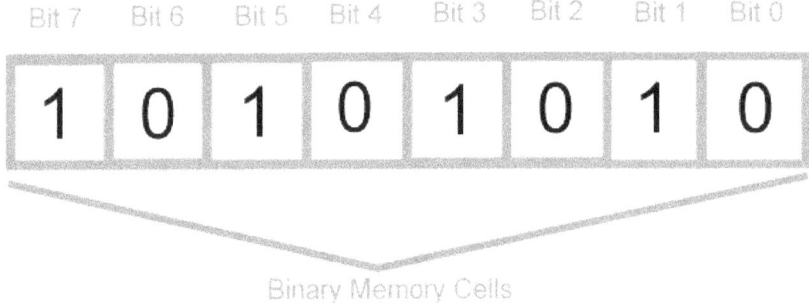

Figure 4-1. 1 Byte or 8 bits

The ov7670 has many different camera registers that are 1 byte or 8 bits long binary memory cells where you can change the values in these cells to set camera features such as resolution, edge enhancement, etc. In the documentation different names are given to different registers based on their purpose in terms of controlling the camera. However, each of these registers is identical in that they are all 8 bits or 1 byte in length and each memory cell can hold a value of either 0 or 1. See Figure 4-2. The CLKRC register controls the camera's internal clock. The SCALING_XSC register controls the horizontal scale factor for the camera's image. The HSTART register controls the horizontal start position of the camera's output frame which may be different depending on resolution.

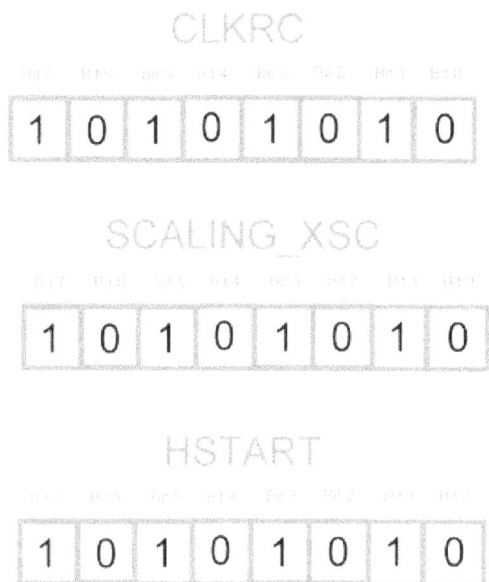

Figure 4-2. ov7670 camera registers

Each bit or group of bits in a register may control a function or functions of the camera. For example, for the CLKRC register bit 7 is reserved and not used, bit 6 controls the use of the external clock, and bit 5 through bit 0 controls the internal clock prescaler. See Figure 4-3.

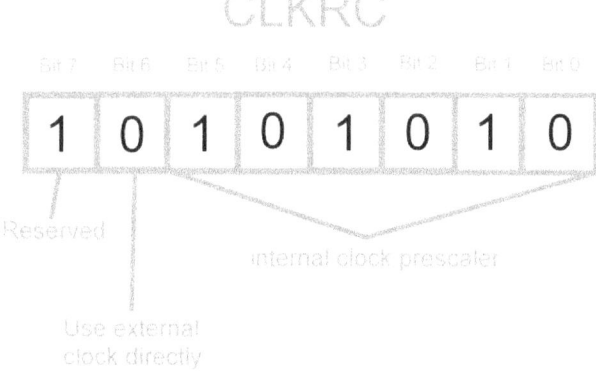

Figure 4-3. The CLKRC register

In order to fully understand the ov7670 data sheet and how to properly set the camera registers you need to understand how various types of number systems such as decimal, binary and hexadecimal representations operate.

Decimal Numbers (Base 10 Representation)

When we think of numbers we generally think of base 10 numbers with digits from 0 through 9. What this means is that for example the number 7952 can be represented as the sum of 7000 + 900 + 50 + 2. We can further represent each of these components as a number multiplied by the base of 10 raised to a power that associated with the component's position in the number. For example, we can rewrite 7000 as 7 * (10*10*10) or 7 * 10^3. The exponent 3 is also the position of "7" in the number. The position is zero based and the exponent of the first component is 0. See Figure 4-4.

So the value of the number 7952 with a base of 10 can calculated by the sum of:

7 * (10*10*10) => 7 * 10^3 => 7000

9 * (10*10)   => 9 * 10^2 => 900

5 * (10)   => 5 * 10^1 => 50

2 * (1)   => 2 * 10^0 => 2

If you add all the numbers you get 7000 + 900 + 50 + 2 which adds up to 7952 in base 10 which is actually what you started with. The point of this example is to lay the groundwork for calculating the base 10 values of numbers that use base 2 (binary) and base 16 (hexadecimal) representation.

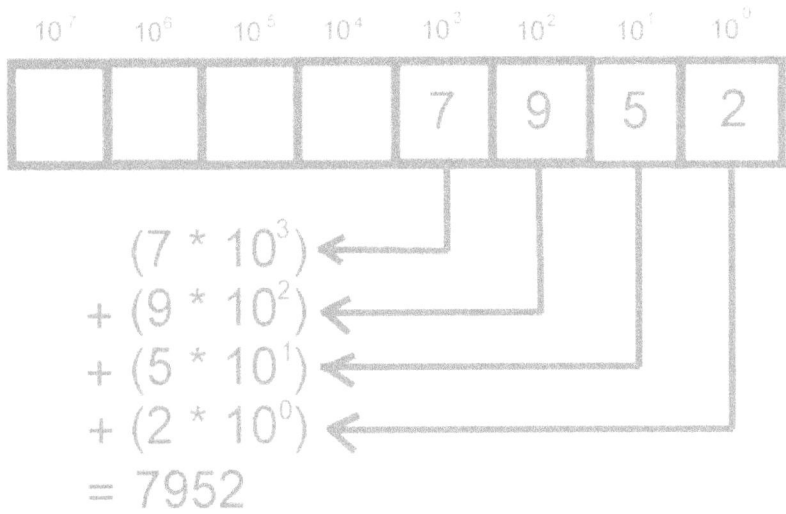

Figure 4-4. Base 10 decimal number

Binary Numbers (Base 2 Representation)

Base 2 numbers or binary numbers have digits consisting of 0 and 1. The decimal (base 10) value for a binary number is calculated by a method similar to that in the previous section. For example, to calculate the decimal value of the binary number 1111 multiply each component of the number by the base 2 raised to the power of the position of that component then sum all the components together. Positions are zero based such that the first position is 0.

1000 => (1 * 2^3) => 8

100 => (1 * 2^2) => 4

10   => (1 * 2^1) => 2

1   => (1 * 2^0) => 1

The individual components of the binary number are then added together to get 8 + 4 + 2 + 1 = 15 in base 10 decimal. See Figure 4-5.

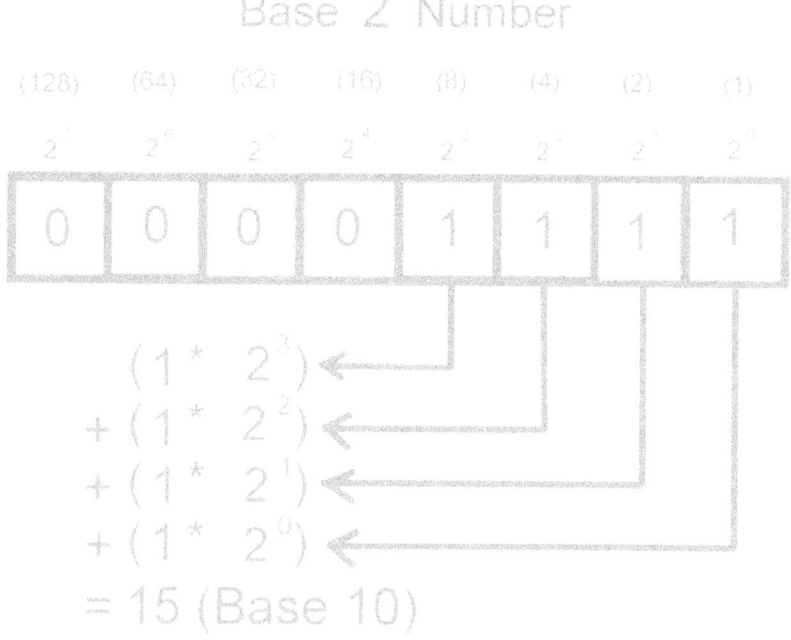

Figure 4-5. Base 2 binary number

Hexadecimal Numbers (Base 16 Representation)

Base 16 numbers or hex numbers consist of digits 0 through 9 and letters A through F. The hex digits 0 through 9 are the same as the decimal digits of 0 though 9. However, the hex digits of A through F correspond in value to 10 through 15 in base 10 or decimal. See Figure 4-6.

The hex number B65F can be converted to decimal using the same method as discussed previously for binary numbers. Each component of the hex number is multiplied by 16 to the power of the component's position in the number and then all these components are added together to get the final base 10 value. The first position being considered 0. Thus, to convert B65F to a decimal number we can do the following:

B000 => B * 16^3 => 11 * 4096 => 45056

600 => 6 * 16^2 => 6 * 256 => 1536

50   => 5 * 16^1 => 5 * 16   => 80

F   => F * 16^0 => 15 * 1   => 15

So adding all these components 45056 + 1536 + 80 + 15 = 46,687 (Base 10). See Figure 4-7 for a more detailed graphic of the above.

Decimal Digit	Hex Digit
0	0
1	1
2	2
3	3
4	4
5	5
6	6
7	7
8	8
9	9
10	A
11	B
12	C
13	D
14	E
15	F

Figure 4-6. Decimal and Hex Digits

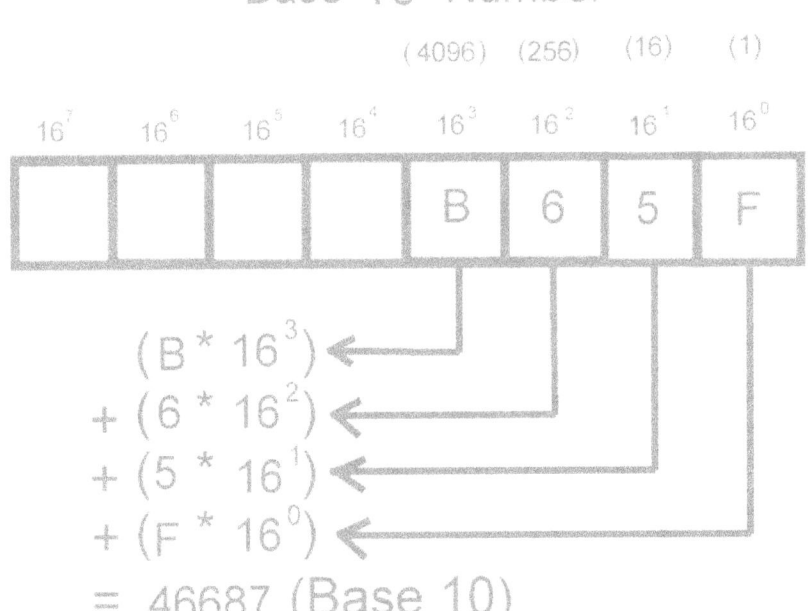

Figure 4-7. Base 16 hex number

Converting a Binary Number (Base 2) to a Hex Number (Base 16)

In this section we will go through an example of how to convert a binary number (base 2) to a hexadecimal number (base 16). The method of doing this is splitting up the binary number into 4 bit groups, converting these groups to hex numbers, and then combining the these hex numbers to form the final hexadecimal number that represents the original binary number.

First we start with the binary number such as 11010101. See Figure 4-8.

Figure 4-8. Binary number 11010101

Next, we need to break the binary number into 4 bit groups such as 1101 and 0101. Each of these groups will be converted to a hex number. See Figure 4-9.

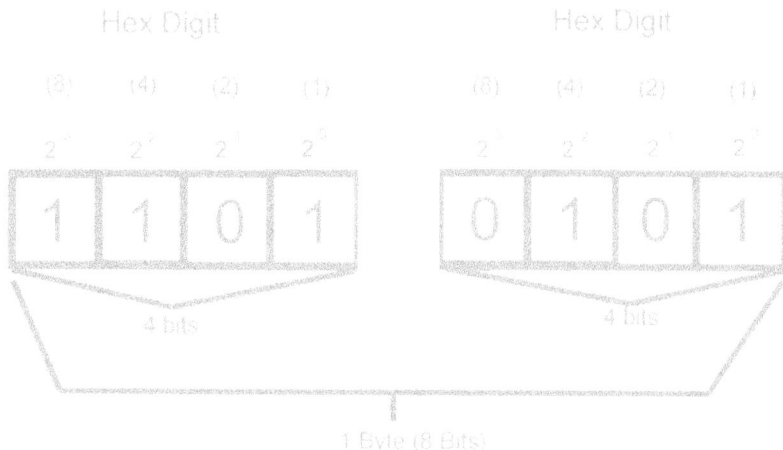

Figure 4-9. Binary to hex conversion

Next, we convert the 4 bit group consisting of 1101 into a single hex digit. To do this we take the components of the binary number and multiply them by 2 raised to the power of their zero based positions and then add these results together and use the table in Figure 4-6 to convert the decimal value into a hex value. For the binary number 1101 the corresponding hex digital is D. See Figure 4-10.

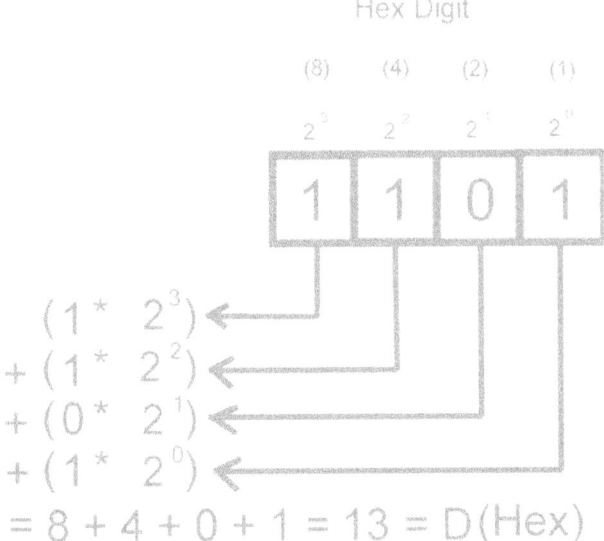

Figure 4-10. Converting the first group of 4 bits of the binary number to the first component of the hex number

Then we convert the other group of 4 bits which is 0101 into a hex number which is 5. We use the same method as for the first group of digits. The key to remember here is that each group of 4 bits from the original binary number is treated like a separate number so that the leftmost digit's value is 2^3 followed by 2^2, 2^1, and 2^0. These translate into the decimal values 8, 4, 2, 1 which are indicated on Figure 4-11.

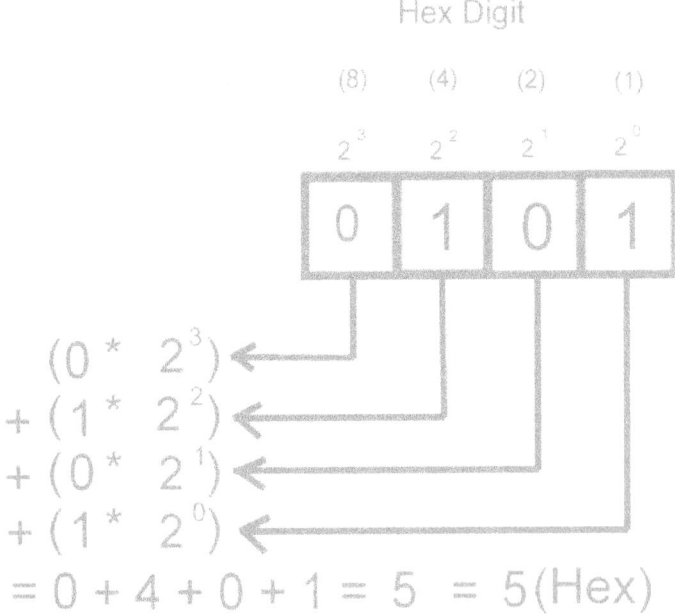

Figure 4-11. Converting the second component of the binary number to the second component of the hex number

The final step is to combine the hex values from each of the 4 bit groups into one single group of digits which is D5. See Figure 4-12 for a summary of the conversion process.

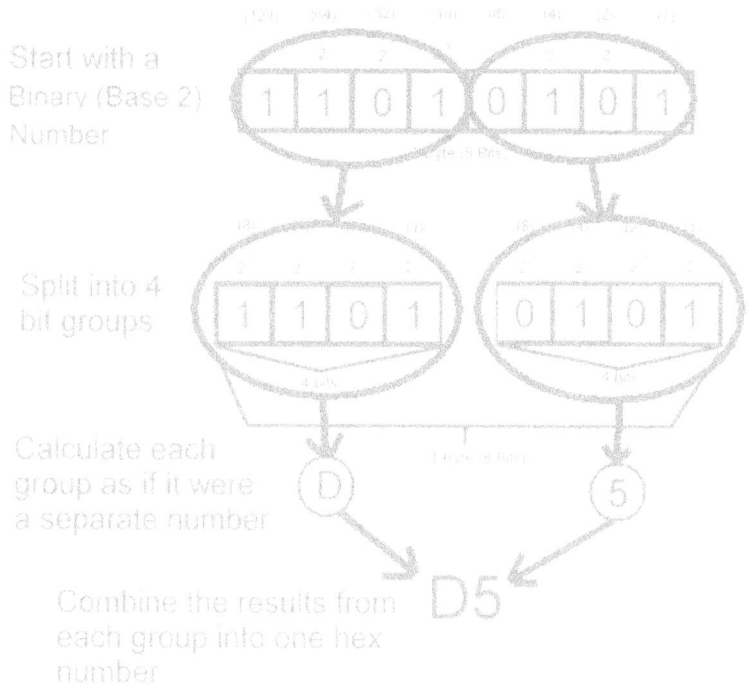

Figure 4-12. Summary of converting a binary number to hex

Converting a Hexadecimal Number (Base 16) to a Binary Number (Base 2)

To convert a hex number (base 16) into a binary number (base 2) you first need to convert each hex digit into a separate group of 4 binary digits that represent that hex digit. The final step would be to combine all the groups of 4 binary digits into one binary number.

For example, the hex number D5 can be converted to binary by first writing each of its components or digits in binary and then combining these two 4 bit binary numbers into one final larger 8 bit binary number. The hex component of D is converted to the 4 bit binary number of 1101. The hex component of 5 is converted to the 4 bit binary number of 0101. The final binary number is formed by combining the binary numbers for D and 5 to get 11010101. See Figure 4-13.

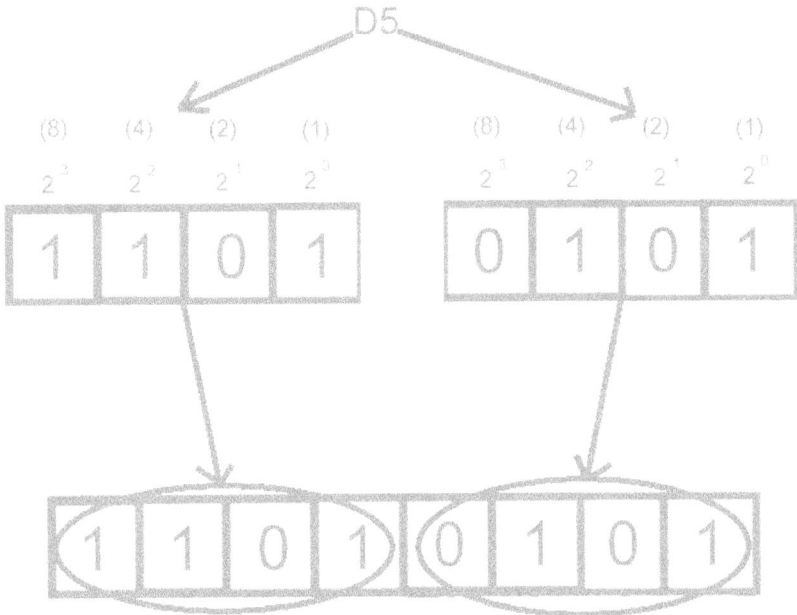

Figure 4-13. Converting a hex number into a binary number

Hands On Example: Setting Registers on the OV7670 Camera

In this hands on example, I demonstrate the importance of what was discussed previously in this chapter in terms of converting a binary number to a hexadecimal number and vice versa. I discuss the COM17 camera register and show you what binary numbers are needed to control certain features of the camera and how these binary numbers can be converted to hexadecimal numbers to be used for setting the actual register. Hexadecimal numbers require less digits than the same number expressed in binary form. Thus, when a human has to actually enter a number to set the value of a camera register into the Arduino IDE a hexadecimal number is preferred because it reduces the chance of data entry error. In addition, the camera's data sheet expresses default register values in hexadecimal so it would be good to know what camera functions are enabled by default when the camera is powered up and to do that you need to know how to convert hex numbers into binary numbers.

The COM17 Camera Register

The COM17 ov7670 camera register controls the camera' automatic exposure controller window size, and the Digital Signal Processor's color bar test enabled status used for testing the camera's output. Bit 7 and bit 6 control the AEC window size. The default setting is 00 which is fine so we do not need to change these bits. Bit 5 and bit 4 are reserved so we can ignore these and leave them at their default settings of 00. Bit 3 is a key bit that enables the color bar test mode of the Digital Signal Processor or DSP. Set this bit to 1 to enable the color test and 0 to disable it. Bits 2, 1, and 0 are reserved so we can leave these at their default values of 000. See Figure 4-14.

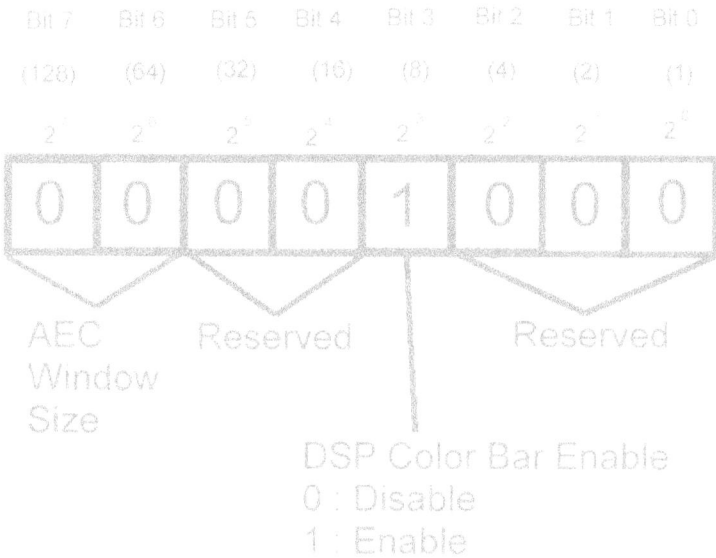

Figure 4-14. Com17 Register

Setting the DSP Color Bar Test Enable

The color bar test is a series of vertical color bars consisting of white, yellow, light blue, green, purple, red, blue, and black as seen from left to right. When the color bar test is enabled normal camera image capturing is disabled and the only output is the color bar test image. This test is useful in determining if the camera is set up correctly and producing the correct color image. See Figure 4-15.

> Important Note: An extremely important thing to note here is that in order to get the correct colors from the ov7670 camera you will need to set a camera register that in the official documentation is listed as being "reserved" but actually is needed to correctly display colors. For example, red colors will appear green in the photo that is taken by the camera. We will discuss this more in depth later in the book and show you exactly how to do it in a "hands on example".

Figure 4-15. The color bar test

In order to enable the color bar test you need to set the COM17 register to the hex values of 08. In order to disable the color bar test you can set the COM17 register to the hex value of 00. Remember that each hex number represents 4 binary bits and thus you can easily get the hex value you need from Figure 4-14.

In terms of implementing these values into Arduino C/C++ source code we can declare defines that represent a value for disabling the color bar test while keeping the AEC window setting set to normal such as:

#define COM17_VALUE_AEC_NORMAL_NO_COLOR_BAR   0x00

We can also declare a define to represent the value for enabling the color bar test while keeping the AEC window setting set to normal such as:

#define COM17_VALUE_AEC_NORMAL_COLOR_BAR   0x08   // Activate Color Bar for DSP

The "0x" in front of the number identifies this number as a hexadecimal number value. We will go into further detail on how to actually set the register value in code in a later "hands on example" in this book. Here, we just are showing you how you change the values in a camera's registers in order to change a certain property in this case it was displaying the color bar test.

Boolean Variables, Logic and Truth Tables

A boolean variable is a variable that can take on one of two values such as 0 or 1. The digital output pins of the Arduino can output electrical signals that represent boolean variables. You can use the built in Arduino function digitalWrite() to output a HIGH or LOW value representing a boolean variable.

For example, to output a HIGH value which represents 1 or true to the pin RRST you would write code such as:

digitalWrite(RRST, HIGH);

To output a LOW value which represents 0 or false to pin RRST you would write code such as:

digitalWrite(RRST, LOW);

Boolean logic applies to boolean variables and can be used in digital circuits to control a boolean output based on two or more boolean inputs. The basic boolean logic operations are AND, OR, and NOT. In order for the output of the AND operator to be true or 1 all of the inputs must be true or 1. See Figure 4-16 for the boolean truth table for AND which lists all the possible inputs and output combinations for the AND operator using two boolean input variables.

Pin X	Pin Y	Pin X AND Pin Y
0	0	0
0	1	0
1	0	0
1	1	1

Figure 4-16. Boolean logic AND

The result of an OR boolean operation is true if any of the inputs are true or 1. See Figure 4-17 for a table of all the possible input/output combinations and results for the OR operation on two binary inputs.

Pin X	Pin Y	Pin X OR Pin Y
0	0	0
0	1	1
1	0	1
1	1	1

Figure 4-17. Boolean logic OR

The result of a NOT boolean operation on a boolean input is the opposite of the input. See the table in Figure 4-18.

Pin X	NOT Pin X
0	1
1	0

Figure 4-18. Boolean logic NOT

Now, let's give you a better idea of how boolean variables relate to the Arduino. Figure 4-19 gives you a graphic representation of the electrical signals which are actually voltage levels that are output by the Arduino that correspond to the AND truth table in Figure 4-16.

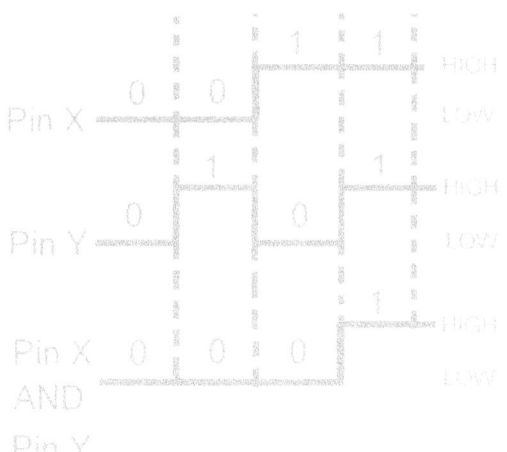

Figure 4-19. Boolean logic AND voltage output for Arduino

Figure 4-20 gives the electrical signals that correspond to the boolean truth table for the OR operator that is listed in Figure 4-17.

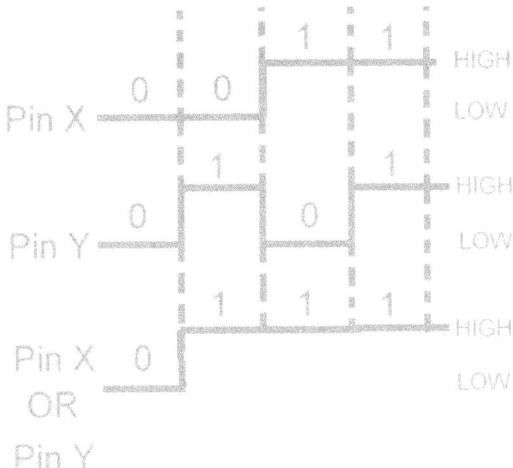

Figure 4-20. Boolean logic OR voltage output for Arduino

The Clock Pulse

The clock pulse is a binary series of HIGH and LOW voltage values that are output from the Arduino and used to drive the state of the camera or other digital device. That is the pulses cause the camera or device to perform specific actions according to its design. For example, in order to read in the data from the camera's memory you need to read the image data one byte at a time until all the bytes are read. In order to read this you need to send a clock pulse to the camera's RCLK input that will be generated from the Arduino. See Figure 4-21.

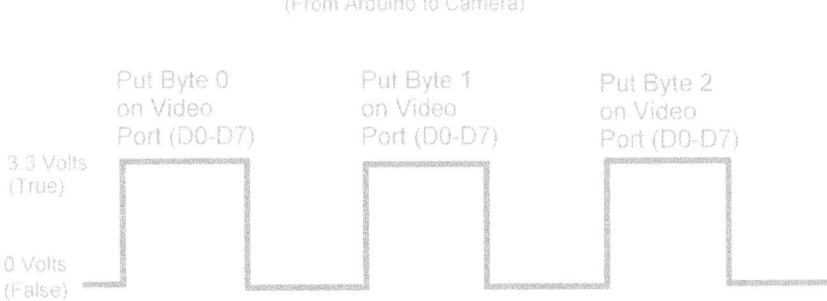

Figure 4-21. RCLK clock pulse output from Arduino to camera

The way I generate a clock pulse to drive the RCLK input is by:

1. Calling the digitalWrite() function to set the desired output pin HIGH

2. Calling the delayMicroseconds() function to wait a specified number microseconds. All this time the output pin that is connected to the RCLK input on the camera is HIGH.

3. Calling the digitalWrite() function and setting the output pin connected to RCLK to LOW.

4. Calling the delayMicroseconds() function and pausing program execution for a specified time. This generates the LOW part of the clock pulse.

See Listing 4-1 for the Arduino code

Listing 4-1. The PulsePin() function

void PulsePin(int PinNumber, int DurationMicroSecs)

{

  digitalWrite(PinNumber, HIGH);   // Sets the pin on

  delayMicroseconds(DurationMicroSecs);   // Pauses for DurationMicroSecs microseconds

  digitalWrite(PinNumber, LOW);   // Sets the pin off

  delayMicroseconds(DurationMicroSecs);   // Pauses for DurationMicroSecs microseconds

}

RCLK is defined as pin 26 using a #define statement

#define RCLK 26   // Output FIFO buffer output clock

Pulsing the actual RCLK pin is done calling the PulsePin() function such as:

PulsePin(RCLK, 1);

Reading Schematics

In order to understand the schematics for the ov7670 camera or digital devices in general you should know the standard symbols for logic gates that implement the boolean logic discussed previously. In this section we will discuss the schematic symbols for the AND, OR and NOT gates. In addition we will discuss the NAND, and NOR gates which are a combination of the AND, and OR gates with a NOT gate.

And Gates

The schematic symbol for an AND gate that implements the AND boolean function is shown in Figure 4-22.

Figure 4-22. And gates

Or Gates

The schematic symbol for the OR gate that implements the OR boolean logic is shown in Figure 4-23.

Figure 4-23. Or gates

Not (Inverter) Gates

The schematic symbol for the NOT gate that implements the NOT boolean logic is shown in Figure 4-24.

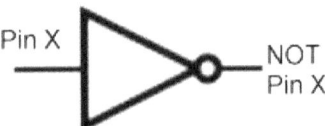

Figure 4-24. Not (inverter) gate

NAND Gates

The schematic symbol for the NAND gate that performs the boolean AND function followed by the boolean NOT function is shown in Figure 4-25. The Texas Instruments SN74LVC1G00 NAND gate is the actual type of gate used in the ov7670 camera according to the publicly available schematics.

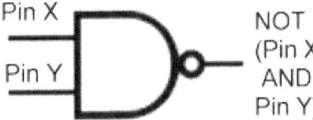

Figure 4-25. NOT AND gate

NOR Gates

The schematic symbol for the NOR gate that performs the boolean OR function followed by the boolean NOT function is shown in Figure 4-26.

Figure 4-26. NOT OR gate

Design Overview for the OV7670 Camera with FIFO Memory

In this section I cover the basic operation of the ov7670 camera from a chip level. I start with discussing the publicly available documentation for the ov7670. I then discuss the main camera chip and the FIFO frame buffer memory chip. I then discuss how these chips are connected using the schematic of the ov7670 camera system. Then, I discuss how the main camera chip outputs a video frame using timing diagrams. Finally, I discuss how the FIFO can capture the video image data from the main camera chip and how you can read the image data from the FIFO.

Publicly Available Documentation for the ov7670

This section describes the publicly available documentation relating to the ov7670 camera including the official documentation from Omnivision, the maker of the ov7670 camera. I recommend that you download the following pdf files which can be read using the Adobe Acrobat Reader which is free and can be downloaded from Adobe's web site. You can also do a google search for these documents. This book is not meant to be a reference manual but a quick start guide that is designed for you to get quickly started developing camera based applications using the Arduino and ov7670. However, the information you learn in this book will also help you develop camera applications for cameras other than the ov7670 as well.

- Omnivision ov7670/ov7171 Advanced Information Preliminary Datasheet (Version 1.4 August 21, 2006)

    http://www.electronicaestudio.com/docs/sht001.pdf

- ov7670/ov7171 Implementation Guide (Version 1.0 September 2,2005)

    https://github.com/dalmirdasilva/ArduinoCamera/blob/master/CameraAL422B/datasheet/OV7670%20Implementation%20Guide%20%28V1.0%29.pdf

- ov7670 Software Application Note

    https://github.com/luckasfb/Development_Documents/blob/master/MTK-Mediatek-Alps-Documents/OV7670%20software%20application%20note.pdf

- AverLogic AL422B Data Sheets (Revision V1.01)

    http://www.frc.ri.cmu.edu/projects/buzzard/mve/HWSpecs-1/Documentation/AL422B_Data_Sheets.pdf

- Omnivision ov7670 FIFO camera schematic

    http://www.beyondlogic.org/pdf/OV7670_FIFO_SCH_V1.pdf

Camera Pin Input/Outputs

The camera input and output pins on the ov7670 are shown below in Figure 4-27.

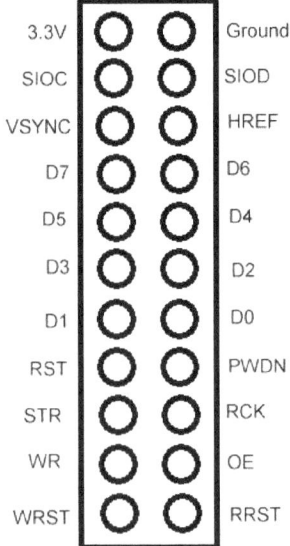

Figure 4-27. Camera pin inputs and outputs

- 3.3v – Input for the 3.3v output pin on the Aruduino

- Ground – Ground connection for the camera to be connected to ground on Arduino

- SIOC – I2C interface connection to the Arduino's SCL clock line.

- SIOD – I2C interface connection to the Arduino's SDA data line.

- VSYNC – Output from camera that marks the start or stop of an output of a single image frame.

- HREF – True if a row of an image is being output to the video port.

- D7 – D0 – Video output port, 1 byte or 8 bits wide. Image data is read from these pins one byte at a time.

- RST – Input that can be used to reset the camera.

- PWDN – Input that can be used to put the camera into power down mode.

- STR – Camera strobe output that can be used to turn on an L.E.D. light while the camera takes an image.

- RCK – Input for a clock pulse that is used to read in data from the video port one byte at a time. One clock cycle corresponding to one byte.

- WR – Write enable input for enabling the writing of data to the camera's frame buffer memory. True if you want image data from the camera to be written to the FIFO frame buffer memory. False if the image data is not to be written to the camera's memory.

- OE – Input pin that controls the camera's output enable for the video port that is used to determine if the data on the pins D7 – D0 are valid.

- WRST – Input pin that is used to reset the frame buffer memory's write pointer so that image data will be written at the start of the image or frame.

- RRST – Input pin that is used to reset the frame buffer memory's read pointer so that image data is read and sent to the video output port starting at the beginning of the image or frame.

Main Camera Chip Overview

This section discusses the main camera chip that captures, processes the image, and sends it out to the frame buffer memory where it can then be read by the Arduino. See Figure 4-28.

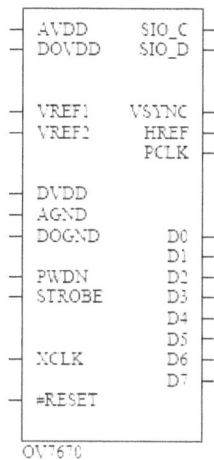

Figure 4-28. Main camera chip

- AVDD – Analog power supply

- SIO_C – SCCB serial interface clock input. (compatible with I2C and SCL)

- DOVDD – Digital power supply for Input/Output

- SIO_D – SCCB serial interface data input and output. (compatible with I2C and SDA)

- VREF1 – Reference voltage.

- VSYNC – Vertical sync output.

- VREF2 – Reference voltage.

- HREF – HREF signal output.

- PCLK – Pixel clock output.

- DVDD – Power supply for the digital logic core

- AGND – Analog ground.

- DOGND – Digital ground.

- D0 – D7 – Video output port that is 8 bits wide and used to output image data.

- PWDN – Power down enable disable mode. 0 is normal mode. 1 is power down mode.

- STROBE – L.E.D. or strobe control output

- XCLK – System clock input

- #RESET – Clears all registers and resets them to their default values. 0 is the reset mode and 1 is the normal mode.

FIFO Frame Buffer Field Memory Overview

This section discusses the camera's frame buffer memory chip. See Figure 4-29.

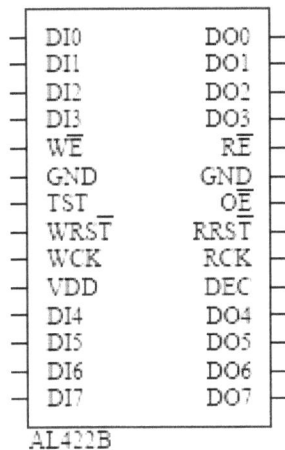

Figure 4-29. AverLogic AL422B FIFO frame buffer memory chip

- DI0 - DI7 – Data input. Data is input on the rising edge of the cycle of WCK when /WE is pulled low (enabled).

- DO0 – DO7 – Data output. Data output is synchronized with the RCK clock. Data is obtained at the rising edge of the RCK clock when /RE is pulled low. The access time is defined from the rising edge of the RCK cycle.

- /WE – Write enable that is active low. /WE controls the enabling/disabling of the data input. When /WE is pulled low, input data is acquired at the rising edge of the WCK cycle. When /WE is pulled high, the memory does not accept data input. The write address pointer is stopped at the current position. /WE signal is fetched at the rising edge of the WCK cycle.

- GND – Ground.

- TST – Test pin. For testing purpose only. It should be pulled low for normal applications.

- /WRST – Write reset that is active low. This reset signal initializes the write address to 0, and is fetched at the rising edge of the WCK input cycle.

- WCK – Write clock. The write data input is synchronized with this clock. Write data is input at the rising edge of the WCK cycle when /WE is pulled low (enabled). The internal write address pointer is incremented automatically with this clock input.

- VDD – 5 volts or 3.3 volts.

- /RE – Read enable that is active low. /RE controls the operation of the data output. When /RE is pulled low, output data is provided at the rising edge of the RCK cycle and the internal read address is incremented automatically. /RE signal is fetched at the rising edge of the RCK cycle.

- /OE – Output enable that is active low. /OE controls the enabling/disabling of the data output. When /OE is pulled low, output data is provided at the rising edge of the RCK cycle. When /OE is pulled high, data output is disabled and the output pins remain at high impedance status. /OE signal is fetched at the rising edge of RCK cycle.

- /RRST – Read reset that is active low. This reset signal initializes the read address to 0, and is fetched at the rising edge of the RCK input cycle.

- RCK – Read clock. The read data output is synchronized with this clock. Read data output at the rising edge of the RCK cycle when /OE is pulled low (enabled). The internal read address pointer is incremented with this clock input.

- DEC – Decoupling cap input. Decoupling cap pin, should be connected to a 1mF or 2.2mF capacitor to ground for 5V application. For 3.3V application, the DEC pin can be simply connected to the 3.3V power with regular 0.1mF bypass capacitor.

Overall Camera Schematic

This section covers the overall camera schematic and shows how the main parts of the camera interact with one another. See Figure 4-30.

Important things to note are:

- The camera clock generator creates a clock pulse that drives the XCLK input of the main camera chip.

- The video port output of the main camera chip pins D0-D7 are connected with the video frame buffer FIFO memory's input data pins which are DI0-DI7.

- The PCLK on the main camera chip that is synced with the output of bytes that represent the pixels of the image is connected to the FIFO memory's WCK that is synced with the writing of this pixel data to the memory.

- The HREF output pin on the main camera chip is connected to a NAND gate with a user input write enable signal that will activate writing data to the FIFO memory when the write enable signal is HIGH and the HREF is HIGH. This results in a HIGH value from the AND operation and a LOW value from the NOT operation which activates the /WE input to the FIFO memory which is active LOW. See Figure 4-31 for a closer view of the NAND gate.

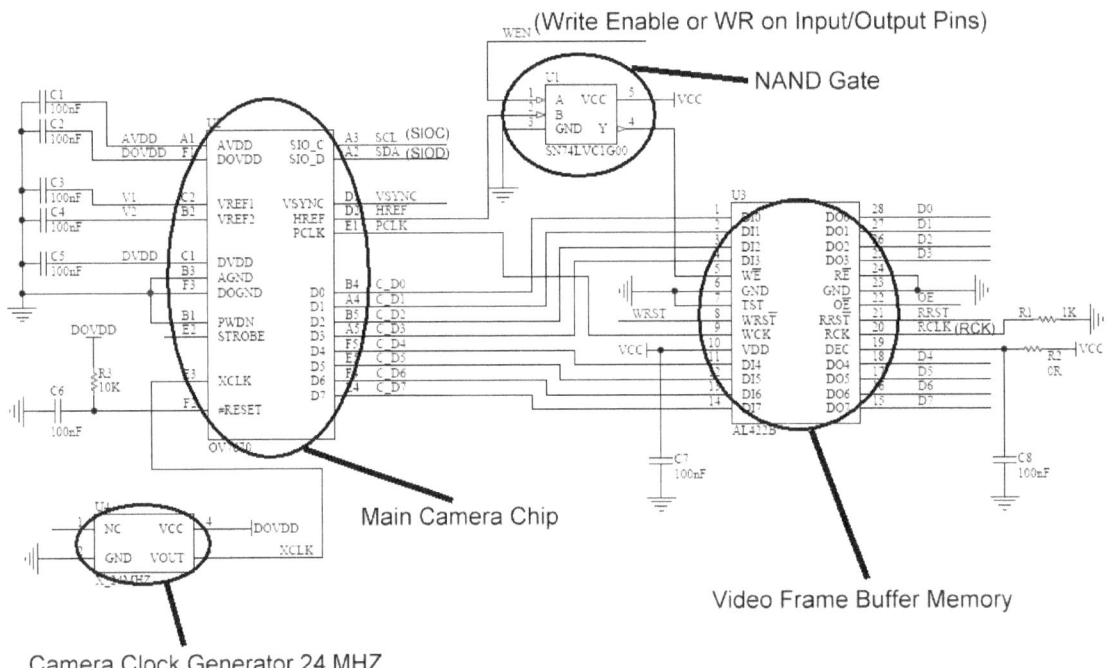

Figure 4-30. Overall Camera Schematic

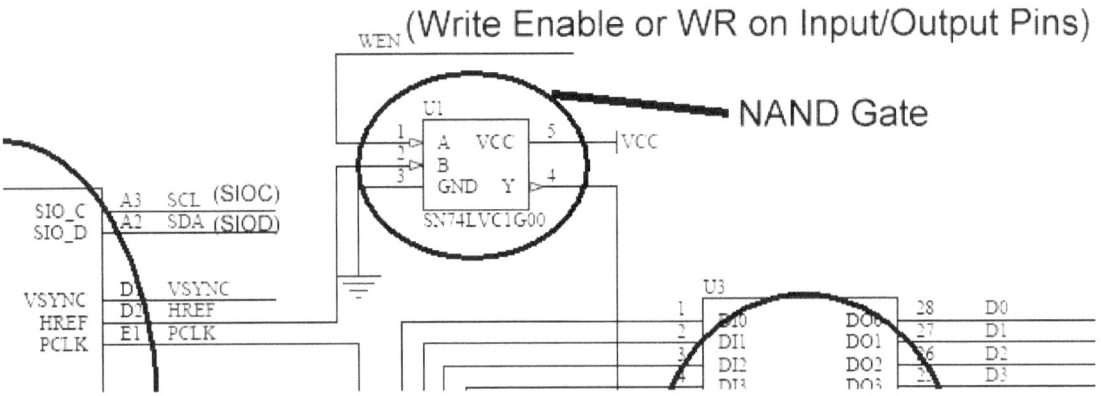

Figure 4-31. Key to writing image data to Memory

Basic Operation

The main camera chip outputs a single video frame image through the 8-bit video port by:

1. Pulsing VSYNC high to indicate the start of a new video frame for output.

2. The HREF goes high when a valid row of pixels from the video frame is put on the output video port pins of D0-D7. This continues for each row from row 0 to the last row for the image size which is row 479 for a VGA screen resolution (640 by 480). In addition, when the HREF is high the PCLK syncs with the output of the bytes for each row in the image. See Figure 4-33.

3. VSYNC pulses high again to indicate the end of the video frame output.

See Figure 4-32 for the timing diagram.

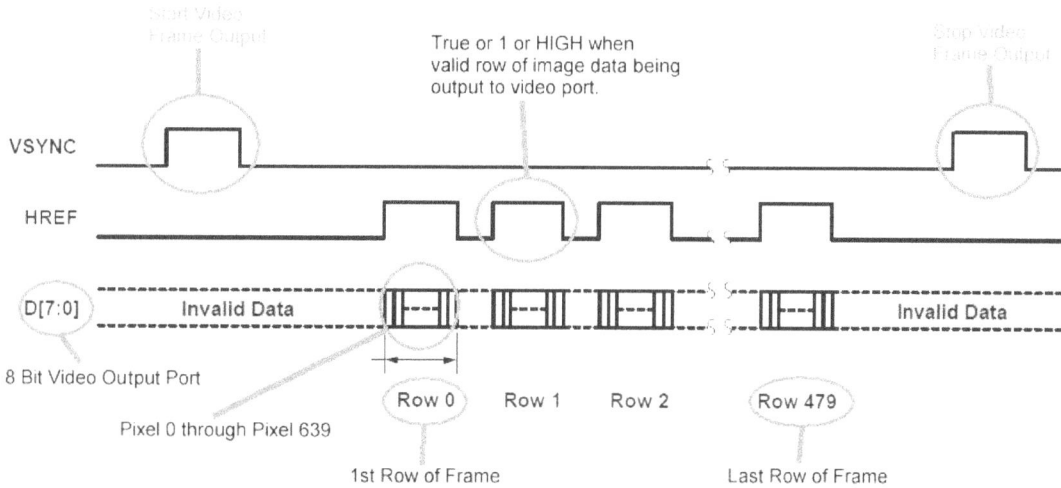

Figure 4-32. Main camera chip VGA image output simplified timing diagram

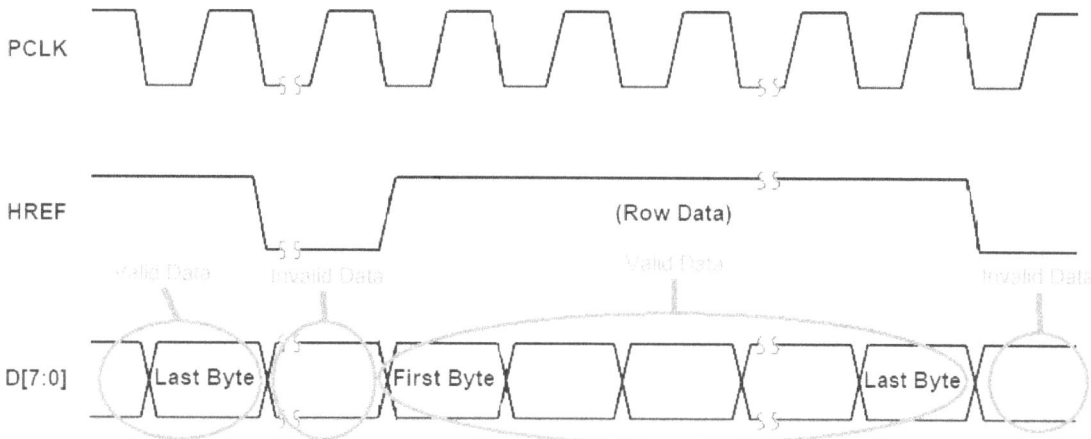

Figure 4-33. The pixel clock

The image output from the main camera chip is then fed into the FIFO video frame buffer memory. From there it is read by the Arduino. The general operation of the frame buffer is given in the functional block diagram in Figure 4-34.

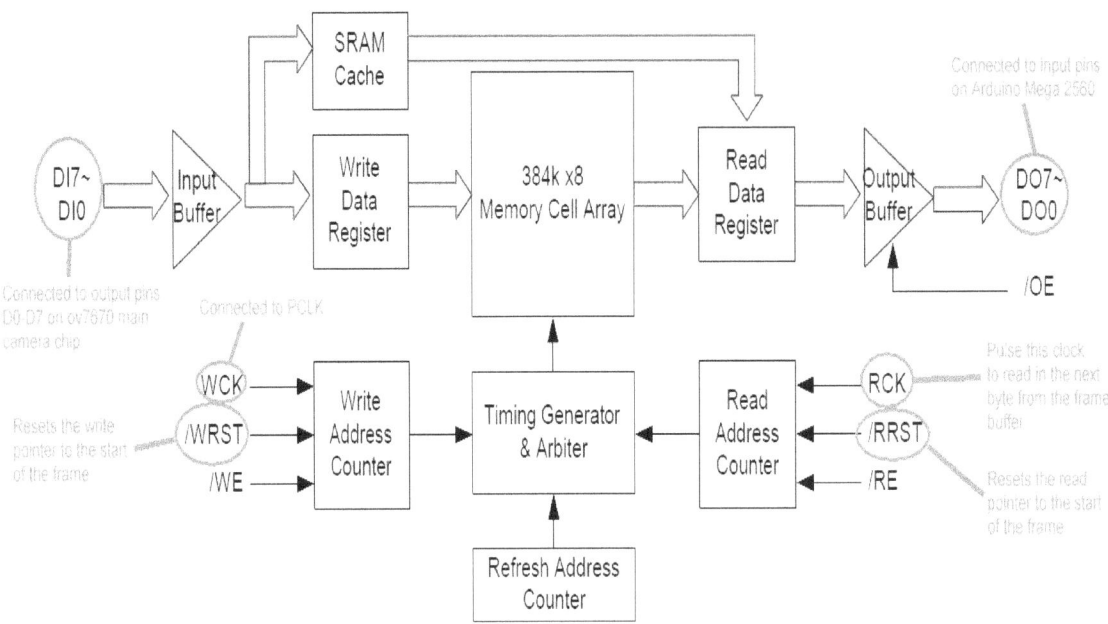

Figure 4-34. Fifo memory operation

In order to capture a video frame being output by the main camera chip you need to:

1. Wait for VSync to pulse to indicate the start of the image

2. Reset Write Pointer to 0 which is the beginning of the frame

3. Set FIFO Write Enable to active (high) so that image can be written to ram

4. Wait for VSync to pulse again to indicate the end of the frame capture

5. Set FIFO Write Enable to nonactive (low) so that no more images can be written to the ram

In order to read the video frame that was captured into the FIFO memory you need to:

1. Set the FIFO read buffer pointer to the start of frame

2. For every byte of data in the image, pulse the read clock RCLK to bring in a new byte of data onto the video output port and then read it into the Arduino.

Once you have the byte of image data in the Arduino's memory you can process this data by writing it to a SD card.

Summary

In this chapter I discussed the basic digital design information that you will need to understand how to develop camera applications. I started with a discussion of how data is stored in a camera. Next, I covered decimal numbers, binary numbers, and hexadecimal numbers. I then covered conversions between these number types. A hands on example followed where I explained how to set the color test bar mode of the camera by setting a camera register. Boolean variables, boolean operations, and boolean truth tables were covered. The clock pulse was discussed and then the schematic symbols for boolean operations such as AND and OR were covered. Finally the design of the ov7670 camera was explained at a chip level.

Chapter 5

Taking Photos with the Omnivision ov7670 Camera – Part 1

In this chapter I cover SD card storage for the Arduino, the Arduino's I2C interface, the Omnivision ov7670 FIFO Camera Image Capture Software, and the ffmpeg image conversion program. I start with a hardware discussion of SD card storage on the Arduino including a discussion of SD card reader input/output pins and an overview of the SD card itself. I discuss the software aspects of the SD card reader and cover reading files, writing files, and deleting files to and from a SD card. Next I cover I2C devices and show you how to connect the I2C interface, initialize, read from, and write to an I2C device such as the ov7670 camera. Then I cover the image capture software that will be used to capture an image from the camera and then to save it to a SD card. Finally, I discuss ffmpeg which is the free image converter software that will be used to convert the image files produced by the camera into PNG files that can be viewed by a paint program or windows explorer.

Overview of SD Card Storage for the Arduino

In this section I discuss the SD card and how to use it with the Arduino. I first cover the SD card reader's output and input pins and how they should be connected to the Arduino Mega 2560. I then show you what the actual SD card looks like and how it should fit into the SD card reader. I then show you have to program the Arduino so that you can write files, read files, and delete files on the SD card.

SD Card Reader Input and Output Pins

The SD card reader input and output pins are shown in Figure 5-1. The input and output pin layouts are from a SD card reader I purchased from Amazon.com. The exact listing was "SD Card Reader Module Slot Socket For Arduino ARM MCU (2pcs) by Exciting $5.50" and you get 2 SD card readers for that price. This SD card reader is for a normal full size SD card.

The input/outputs pins of the SD card reader are:

- GND – Ground connection.

- 3.3 – 3.3 volt input

- 5V – 5 volt input

- SDCS – SD card's Chip Select Pin. This is the pin that the master can use to enable and disable specific devices. When a device's Chip Select (also called the Slave Select) pin is low, it communicates with the master. When it's high, it ignores the master. This allows you to have multiple devices sharing the same MISO, MOSI, and CLK lines.

- MOSI – Master Out Slave In. The master line for sending data to the peripherals.

- SCK – Serial Clock. The clock pulses which synchronize data transmission generated by the master.

- MISO – Master In Slave Out. The slave line for sending data to the master.

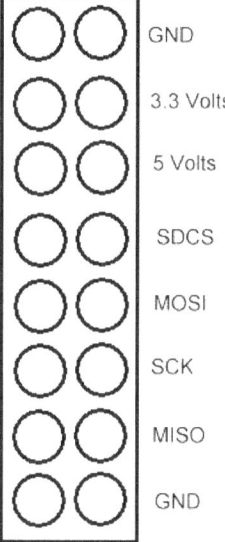

Figure 5-1. SD card reader input/output pins

In terms of the connecting the SD card to the Arduino Mega, this is done by:

- Connecting Pin 50 on the Arduino to the (MISO) pin on the SD card

- Connecting Pin 51 on the Arduino to the (MOSI) pin on the SD card

- Connecting Pin 52 on the Arduino to the (SCK) pin on the SD card.

- Connecting Pin X (User Defined) on the Arduino to the (SDCS) pin on the SD card or connecting Pin 53 on the Arduino (Hardware SS or Slave Select) to the (SDCS) pin on the SD card.

- Connecting the GND pin on the card reader to the GND pin on the Arduino.

- Connecting the 3.3 V pin on the card reader to the 3.3v output on the Arduino.

Important Note: There are two pins for each input/output item for the SD card. I recommend using the pin closest to the pin label since I have seen comments that on some cards the pin that is farthest from the pin label does not work. See Figure 5-2 to see how I set up my SD card.

Figure 5-2. SD card reader connections

The SD card

The SD card itself goes into the metal holder face up and oriented as shown in Figure 5-3. You will need to push the card in until you hear a click. This indicates that the card is fully in and is ready for use. To remove the card press down until you hear another click and then remove the card.

Figure 5-3. SD card reader with card

I purchased my SD card from Amazon.com and according to the listing it was a "SanDisk 8GB Class 4 SDHC Memory Card". This will provide plenty of memory for taking pictures with the camera. See Figure 5-4 for an image of the SanDisk SD card.

Figure 5-4. SDHC 8gb card from Sandisk

Using the SD Card with Arduino

The Arduino comes with a built in library of functions that allow access to the SD card. There are functions that allow the user to write data to the SD card, read data from the SD card, check if a file on the SD card exists, and delete a file from the SD card.

Initializing the SD Card

In order to use the SD library functions you need to include the SD library header in your source code such as:

#include <SD.h>

The hardware slave select pin is pin 53 on the Arduino Mega

const int HardwareSSPin = 53;   // For Arduino Mega

The hardware slave select pin also must be set to be of an output type in order for the SD card library to work. We call the built in pinMode() function to set the hardware slave select pin to be of an output type. This is needed even if it is not used. For example, the actual chip select pin is other than 53.

pinMode(HardwareSSPin, OUTPUT);   // change this to 53 on a mega

Next, we define a chip select pin as pin 48. We could have easily defined the chip select pin as another digital pin if we wished.

const int chipSelect = 48;

Then, we need to initialize the SD card and SD card library by calling the SD.begin() function. Here we have the option of changing the default chip select pin of 53 (on the Mega) which is called the hardware SS pin to a pin specified in the number to sent the begin() function.

For example, we can initialize the SD card with the chip select pin being set to 48 by the following code. If the initialization succeeds then a true value is returned otherwise a false value is returned.

```
if (!SD.begin(chipSelect))

{

    // SD card initialization failed.

    return;

}

else

{

    // SD card initialization successful

}
```

Writing Files to the SD Card

You can also write files to the SD card using the built in SD card library. In the hands on example presented later in this chapter camera data is read in one byte at a time into the Arduino's memory and immediately written to the SD card.

In order to write a file to a SD card you need to:

1. Declare a variable of type File that represents the file that you will write to.

2. Call the SD.open() function with the parameters for the filename and the parameter "FILE_WRITE" to create a file for reading and writing starting at the end of the file. Set the return value of the function to the variable that you declared as a File in Step 1.

3. If the return value from the SD.open() function is null or 0 then there was an error opening the file. Otherwise the File object is valid.

4. Once you have a valid File object you can write text data to the file by calling the print() function on the object. You can write binary data to the file by calling the object's write() function.

5. Finally, in order to save the data to the SD card you need to call the close() function on the File object.

See Listing 5-1 for an example of writing a file to a SD card.

Listing 5-1. Writing a file to a SD card

```
File TempFile;

byte Data;


TempFile = SD.open("testfile.txt", FILE_WRITE);
```

```
if (TempFile)
{
    TempFile.print("Data to be written"));
    TempFile.write(Data);

    // close the file:
    TempFile.close();
}
else
{
    // Error opening file
}
```

Reading Files from the SD Card

You can read files that you have saved from the SD card.

To read data from a file that has been saved to a SD card and print it to the screen from the Serial Monitor you must:

1. Declare a variable of type File that represents the file to be read in.

2. Call SD.open() with the filename that is to read in. Set the return value of the function to the File variable created in Step 1.

3. If the File variable is null or 0 then the opening of the file failed.

4. If the File variable is valid then you can read in the contents. The File object's available() function can be called to determine if there are any more bytes to read from the file. Use the read() function to actually read in the bytes and print them to screen. You can put the read() function into a while loop and the test part of the loop can test to see if there are more bytes available for reading.

5. After reading all the bytes from the file and processing them call the close() function to close the file.

See Listing 5-2 for an example of how to read in a file from a SD card and print the contents to the screen using the Serial Monitor.

Listing 5-2. Reading files from the SD card.

```
File TempFile;
```

```
// Reads in file and prints it to screen via Serial
TempFile = SD.open("testfile.txt");
if (TempFile)
{
  // read from the file until there's nothing else in it:
  while (TempFile.available())
  {
    Serial.write(TempFile.read());
  }
  // close the file:
  TempFile.close();
}
else
{
  // Error opening file
  Serial.print("Error opening ");
}
```

Remove existing Files from the SD Card

When writing files to the SD card you may need to save over an existing file. However, the open() command only allows you to open a file for writing starting at the end of the file. There is no option to open a file for writing and start at the beginning of the file. So we need to test for the existence of the file we want to write and remove it if it exists.

In order to write over an existing file we need to:

1. Check to see if that file already exists using the SD.exists() function with the input parameter of the filename we want to write.

2. If the SD.exists() function returns true then you need to call the SD.remove() function with the filename as a parameter to remove the file from the SD card.

3. Now you are able to write the new file using the code in Listing 5-1.

See Listing 5-3 for an example of how to test for the existence of a file and then remove it if it does exist.

Listing 5-3. Removing existing files from the SD card

// Check if file already exists and remove it if it does

if (SD.exists("testfile.txt"))

{

  SD.remove("testfile.txt");

}

Overview of Arduino's I2C Interface

This section covers the Arudino's I2C interface. I first cover the input/output pins for the I2C on the Arduino and the ov7670 camera. Then, I show you how to initialize, write to, and read from a I2C device using the Arduino.

I2C Input/Output Interface Pins

The I2C interface consists of two pins the SDA pin which handles data and the SCL pin which provides the clock.

On the Arduino Mega, SDA is digital pin 20 and SCL is 21. See Figure 5-5 where the SDA and SCL pins are circled.

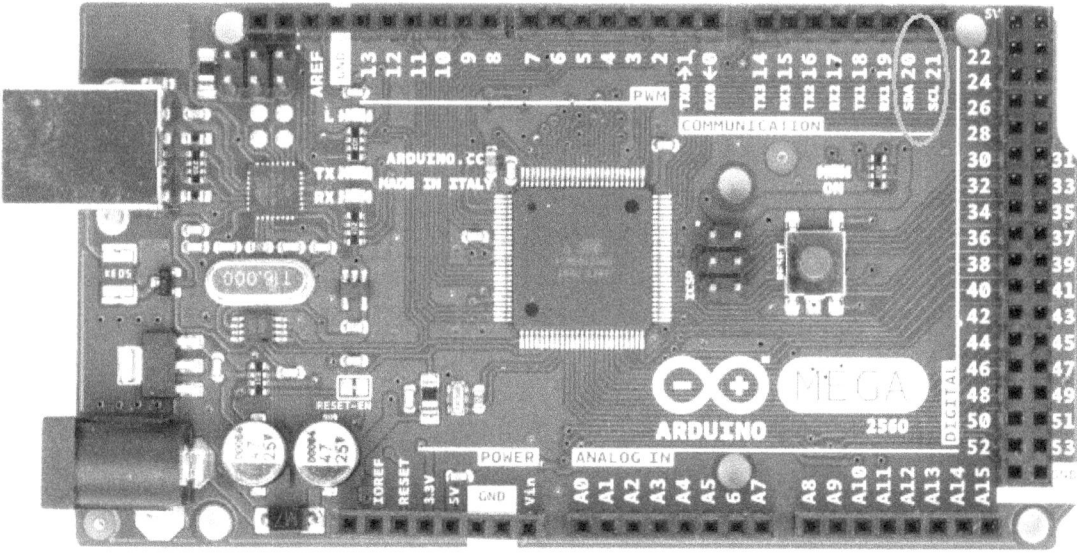

Figure 5-5. Arduino Mega 2560 with SDA and SCL pins circled

On the ov7670 camera the SIOC is hooked to the Arduino's SCL pin. The SIOD is connected to the Arduino's SDA pin. See Figure 5-6 for the location of these pins on the camera.

85

Figure 5-6. ov7670 camera with SIOC and SIOD pins circled

Using an I2C device with Arduino

Arduino has built in support for devices with an I2C bus or compatible bus such as the ov7670 camera. In this section I will show you how to initialize, write data to, and read data from an I2C device.

Initializing an I2C Device

In order to use an I2C device you need to first initialize the I2C bus. You do this by first including the Wire library into your source code. Such as:

#include <Wire.h>

Then you need to call the Wire.begin() function in the setup() function to initialize the Arduino's Wire library such as:

void setup()

{

  Wire.begin();

}

Reading Data from a I2C Device

You can read data from an I2C device by:

1. Calling the Wire.beginTransmission() function with the address of the I2C device.

2. Calling the Wire.write() function with the register address to read from.

3. Calling the Wire.endTransmission() function to send the register address to the I2C device (in this case the camera) and to end the transmission which releases control of the I2C bus.

4. Calling the Wire.requestFrom() function to request one or more bytes from the I2C device.

5. Waiting for the byte to become available by continually calling Wire.available() until a byte becomes available.

6. When a byte become available then read in the byte by calling Wire.read().

See Listing 5-4 for a function that reads and returns a single byte from an I2C device with input address RegisterAddress.

Listing 5-4. The ReadRegisterValue() function

```
byte ReadRegisterValue(int RegisterAddress)
{
  byte data = 0;


  Wire.beginTransmission(OV7670_I2C_ADDRESS);
  Wire.write(RegisterAddress);
  Wire.endTransmission();
  Wire.requestFrom(OV7670_I2C_ADDRESS, 1);
  while(Wire.available() < 1);
  data = Wire.read();


  return data;
}
```

Writing Data to a I2C Device

You can write data to an I2C device by:

1. Calling the Wire.beginTransmission() function with the address of the I2C device.

2. Calling the Wire.write() function with the starting address of the register or registers to write to.

3. Calling the Wire.write(pData, size) function with the array of data in bytes that is to be written and the number of bytes that are to be written.

4. Calling the Wire.endTransmission() function which ends the transmission and sends the bytes to the I2C device. Once completed a stop message is sent and the I2C bus is released.

Listing 5-5 shows how this is done in the hands on example presented later in this chapter.

Listing 5-5. The OV7670Write() function

```
int OV7670Write(int start, const byte *pData, int size)
{
  int n, error;
  Wire.beginTransmission(OV7670_I2C_ADDRESS);
  n = Wire.write(start);      // write the start address
  if (n != 1)
  {
    return (I2C_ERROR_WRITING_START_ADDRESS);
  }
  n = Wire.write(pData, size);  // write data bytes
  if (n != size)
  {
    return (I2C_ERROR_WRITING_DATA);
  }
  error = Wire.endTransmission(true); // release the I2C-bus
  if (error != 0)
  {
    return (error);
  }
  return 0;      // return : no error
}
```

Hands on Example: Testing the I2C Interface with the OV7670 Camera

In this hands on example we are going to run an I2C scanner which detects if there are any I2C devices connected to the Arduino. If there are I2C devices connected then it will print out the device's address. This is a good way to check to see if your ov7670 camera can be detected by the Arduino.

First you will need to download the I2C Scanner program at:

http://playground.arduino.cc/Main/I2cScanner

which is considered the main official site for the Arduino.

The I2C scanner program continually scans the I2C bus for devices by:

1. Calling the Wire.beginTransmission() function with an address in the range of 1 through 127.
2. Calling the Wire.endTransmission() function and testing the return error code.
3. If the return error code is 0 which means there was no error then a valid I2C device has been detected.
4. If a valid I2C device has been detected then print out the device's address to the screen using the Serial Monitor.

See Listing 5-6 for the complete I2C scanner code.

Listing 5-6. I2C device scanner

```
#include <Wire.h>

void setup()
{
  Wire.begin();

  Serial.begin(9600);

  Serial.println("\nI2C Scanner");
}


void loop()
{
  byte error, address;

  int nDevices;
```

```
Serial.println("Scanning...");
nDevices = 0;
for(address = 1; address < 127; address++ )
{
  // The i2c_scanner uses the return value of
  // the Write.endTransmisstion to see if
  // a device did acknowledge to the address.
  Wire.beginTransmission(address);
  error = Wire.endTransmission();

  if (error == 0)
  {
    Serial.print("I2C device found at address 0x");
    if (address<16)
      Serial.print("0");
    Serial.print(address,HEX);
    Serial.println("  !");
    nDevices++;
  }
  else if (error==4)
  {
    Serial.print("Unknow error at address 0x");
    if (address<16)
      Serial.print("0");
    Serial.println(address,HEX);
  }
}
if (nDevices == 0)
```

```
  Serial.println("No I2C devices found\n");

else

  Serial.println("done\n");



  delay(5000);          // wait 5 seconds for next scan
}
```

Using the Scanner Program

1. Connect the Arduino Mega 2560's SDA digital pin 20 to the SIOD pin on the camera.

2. Connect the Arduino Mega 2560's SCL digital pin 21 to the SIOC pin on the camera.

3. Connect the 3.3 volt pin on the Arduino to the 3.3 volt pin on the camera.

4. Connect the GND pin on the Arduino to the GND pin on the camera.

5. Upload the I2C Scanner code to your Arduino.

6. Hit the Serial Monitor button and wait for the test to run.

7. The program should eventually detect the camera that is attached to the I2C bus and report that the address of the camera is 0x21. You will need to use this address to read and write data using the camera.

Overview of the Omnivision ov7670 FIFO Camera Image Capture Software

In this section I discuss in detail the software that I created to capture images from the camera. I first discuss initializing the program and I then cover the main program loop. The initialization involves initializing the camera and SD card. The main program loop consists of processing various user inputs from the Serial Monitor and performing certain actions based on these user inputs.

Initializing the Program

The first thing that is run in an Arduino program is the user defined code located in the setup() function.

In this setup() function the following actions are taken:

1. The serial monitor is initialized so that you will be able to give the camera and SD card commands.

2. The ov7670 camera is initialized and set up for use in taking photos.

3. The SD card is initialized.

The Serial Monitor is initialized by:

1. Calling the Serial.begin() function with 9600 as the rate of communication. Also, make sure you have the Serial Monitor set to 9600 as well which also should be the default.
2. Printing out a text message to the Serial Monitor indicating that the program has started.

See Listing 5-7.

Listing 5-7. Initializing the Serial Monitor

```
// Initialize Serial

Serial.begin(9600);

Serial.println(F("Arduino SERIAL_MONITOR_CONTROLLED_CAMERA ... Using ov7670 Camera"));

Serial.println();
```

Camera Initialization

The camera is initialized by:

1. Calling the Wire.begin() function which initializes the library that handles input and output using the camera's I2C interface
2. The ResetCameraRegisters() function is called which resets all the camera's registers to their default value.
3. The ReadRegisters() function is called which reads and prints out some key camera registers useful for debugging and checking if the camera is working.
4. The SetupCamera() camera function calls the InitializeOV7670Camera() function to prepare the camera for taking photos.

See Listing 5-8.

Listing 5-8. Initializing the camera

```
// Setup the OV7670 Camera for use in taking still photos

Wire.begin();

Serial.println(F("---------------------------- Camera Registers ---------------------------"));

ResetCameraRegisters();

ReadRegisters();

Serial.println(F("------------------------------------------------------------------------"));

SetupCamera();
```

Serial.println(F("FINISHED INITIALIZING CAMERA ..."));

Serial.println();

Serial.println();

The ResetCameraRegisters() function sets all the camera registers by:

1. Calling the ReadRegisterValue() function. This is sometimes needed in order to reliably write a value to a register based on experiments I have done. This is most likely a bug in the ov7670. See Listing 5-4 for a more detailed explanation of this function.

2. Calling the OV7670WriteReg(COM7, COM7_VALUE_RESET) function with COM7 as the register that you want to write to and the COM7_VALUE_RESET value as the value to write as the second parameter.

3. Calling the ParseI2CResult() function which converts the result of the previous I2C operation into an easily readable text string that is printed out.

4. Calling delay(500) which delays the program execution for 500 milliseconds in order to allow time for all the camera registers to reset.

See Listing 5-9.

Listing 5-9. The ResetCameraRegisters() function

```
void ResetCameraRegisters()

{

 // Reset Camera Registers

 // Reading needed to prevent error

 byte data = ReadRegisterValue(COM7);


 int result = OV7670WriteReg(COM7, COM7_VALUE_RESET );

 String sresult = ParseI2CResult(result);

 Serial.println("RESETTING ALL REGISTERS BY SETTING COM7 REGISTER to 0x80: " + sresult);


 // Delay at least 500ms

 delay(500);

}
```

The OV7670WriteReg(int reg, byte data) writes data to a register reg by calling the OV7670Write() function and returns the error state of the operation. See Listing 5-5 for a more detailed explanation of OV7670Write() function. See Listing 5-10 for the code for the OV7670WriteReg() function.

Listing 5-10. OV7670WriteReg() function

```
int OV7670WriteReg(int reg, byte data)
{
  int error;
  error = OV7670Write(reg, &data, 1);
  return (error);
}
```

The ParseI2Cresult() function converts the integer result from an I2C read/write operation into an easily readable text string which is returned to the caller. See Listing 5-11.

Listing 5-11. ParseI2Cresult() function

```
String ParseI2CResult(int result)
{
  String sresult = "";
  switch(result)
  {
    case 0:
      sresult = "I2C Operation OK ...";
      break;

    case I2C_ERROR_WRITING_START_ADDRESS:
      sresult = "I2C_ERROR_WRITING_START_ADDRESS";
      break;

    case I2C_ERROR_WRITING_DATA:
      sresult = "I2C_ERROR_WRITING_DATA";
      break;

    case DATA_TOO_LONG:
      sresult = "DATA_TOO_LONG";
```

      break;

   case NACK_ON_TRANSMIT_OF_ADDRESS:

    sresult = "NACK_ON_TRANSMIT_OF_ADDRESS";

   break;

   case NACK_ON_TRANSMIT_OF_DATA:

    sresult = "NACK_ON_TRANSMIT_OF_DATA";

   break;

   case OTHER_ERROR:

    sresult = "OTHER_ERROR";

   break;

   default:

    sresult = "I2C ERROR TYPE NOT FOUND...";

   break;

  }

  return sresult;

}

The ReadRegisters() function reads in some of the key registers and prints them out to the screen. See Listing 5-12.

Listing 5-12. The ReadRegisters() function

void ReadRegisters()

{

 byte data = 0;

 data = ReadRegisterValue(CLKRC);

 Serial.print(F("CLKRC = "));

```
        Serial.println(data,HEX);


        data = ReadRegisterValue(COM7);
        Serial.print(F("COM7 = "));
        Serial.println(data,HEX);


        data = ReadRegisterValue(COM3);
        Serial.print(F("COM3 = "));
        Serial.println(data,HEX);


        data = ReadRegisterValue(COM14);
        Serial.print(F("COM14 = "));
        Serial.println(data,HEX);


        data = ReadRegisterValue(SCALING_XSC);
        Serial.print(F("SCALING_XSC = "));
        Serial.println(data,HEX);


        data = ReadRegisterValue(SCALING_YSC);
        Serial.print(F("SCALING_YSC = "));
        Serial.println(data,HEX);


        data = ReadRegisterValue(SCALING_DCWCTR);
        Serial.print(F("SCALING_DCWCTR = "));
        Serial.println(data,HEX);


        data = ReadRegisterValue(SCALING_PCLK_DIV);
        Serial.print(F("SCALING_PCLK_DIV = "));
        Serial.println(data,HEX);
```

```
data = ReadRegisterValue(SCALING_PCLK_DELAY);
Serial.print(F("SCALING_PCLK_DELAY = "));
Serial.println(data,HEX);


// default value D
data = ReadRegisterValue(TSLB);
Serial.print(F("TSLB (YUV Order- Higher Bit, Bit[3]) = "));
Serial.println(data,HEX);


// default value 88
data = ReadRegisterValue(COM13);
Serial.print(F("COM13 (YUV Order - Lower Bit, Bit[1]) = "));
Serial.println(data,HEX);


data = ReadRegisterValue(COM17);
Serial.print(F("COM17 (DSP Color Bar Selection) = "));
Serial.println(data,HEX);


data = ReadRegisterValue(COM4);
Serial.print(F("COM4 (works with COM 17) = "));
Serial.println(data,HEX);


data = ReadRegisterValue(COM15);
Serial.print(F("COM15 (COLOR FORMAT SELECTION) = "));
Serial.println(data,HEX);


data = ReadRegisterValue(COM11);
Serial.print(F("COM11 (Night Mode) = "));
```

```
Serial.println(data,HEX);


data = ReadRegisterValue(COM8);
Serial.print(F("COM8 (Color Control, AWB) = "));
Serial.println(data,HEX);


data = ReadRegisterValue(HAECC7);
Serial.print(F("HAECC7 (AEC Algorithm Selection) = "));
Serial.println(data,HEX);


data = ReadRegisterValue(GFIX);
Serial.print(F("GFIX = "));
Serial.println(data,HEX);


// Window Output
data = ReadRegisterValue(HSTART);
Serial.print(F("HSTART = "));
Serial.println(data,HEX);


data = ReadRegisterValue(HSTOP);
Serial.print(F("HSTOP = "));
Serial.println(data,HEX);


data = ReadRegisterValue(HREF);
Serial.print(F("HREF = "));
Serial.println(data,HEX);


data = ReadRegisterValue(VSTRT);
Serial.print(F("VSTRT = "));
```

```
    Serial.println(data,HEX);


    data = ReadRegisterValue(VSTOP);

    Serial.print(F("VSTOP = "));

    Serial.println(data,HEX);


    data = ReadRegisterValue(VREF);

    Serial.print(F("VREF = "));

    Serial.println(data,HEX);
}
```

The InitializeOV7670Camera() sets up the camera so that it can be controlled by the Arduino.

The function does the following:

1. Sets the WRST pin on the Arduino for digital output.

2. Sets the RRST pin on the Arduino for digital output.

3. Sets the WEN pin on the Arduino that is connected to the WR pin on the camera for digital output.

4. Sets the VSYNC pin on the Arduino for digital input. Input from this pin is used for determining when an image should be captured.

5. Sets the RCLK pin on the Arduino for digital output.

6. Sets the D0-D7 pins on the Arduino for digital input. These are the pins that will read in the image data one byte at a time.

7. Calls the delay() function to made sure the camera has had enough time to fully power up.

8. Calls the PulseLowEnabledPin(WRST, DurationMicroSecs) function which sends a low pulse that resets the write buffer pointer on the FIFO.

9. Calls the digitalWrite(RRST, LOW) function to write a LOW value to reset the read FIFO pointer.

10. Calls the PulsePin(RCLK, DurationMicroSecs) function that pulses the RCLK pin so that the camera will recognize that the RRST value is LOW and resets the FIFO read pointer.

11. Calls the digitalWrite(RRST, HIGH) function to restore the RRST value of HIGH or false since this pin is low enabled (low is true , high is false). Thus the read pointer will not reset which RCLK is pulsed.

See Listing 5-13.

Listing 5-13. The InitializeOV7670Camera() function

```
void InitializeOV7670Camera()
{
  Serial.println(F("Initializing OV7670 Camera ..."));

  //Set WRST to 0 and RRST to 0 , 0.1ms after power on.
  int DurationMicroSecs = 1;

  // Set mode for pins wither input or output
  pinMode(WRST , OUTPUT);
  pinMode(RRST , OUTPUT);
  pinMode(WEN  , OUTPUT);
  pinMode(VSYNC, INPUT);
  pinMode(RCLK , OUTPUT);

  // FIFO Ram output pins
  pinMode(DO7 , INPUT);
  pinMode(DO6 , INPUT);
  pinMode(DO5 , INPUT);
  pinMode(DO4 , INPUT);
  pinMode(DO3 , INPUT);
  pinMode(DO2 , INPUT);
  pinMode(DO1 , INPUT);
  pinMode(DO0 , INPUT);

  // Delay 1 ms
  delay(1);

  PulseLowEnabledPin(WRST, DurationMicroSecs);
```

```
// Need to clock the fifo manually to get it to reset

digitalWrite(RRST, LOW);

PulsePin(RCLK, DurationMicroSecs);

PulsePin(RCLK, DurationMicroSecs);

digitalWrite(RRST, HIGH);

}
```

The PulseLowEnabledPin() function takes a pin number and duration in microseconds as input and outputs a pulse of LOW and HIGH values on the input pin number of the specified duration. The idea is to bring the low enabled pin LOW or true for a short period of time to activate it and then turn it off by bringing it back to HIGH or false.

> Note: Remember that a low enabled pin is one in which a LOW input activates the pin and a HIGH input deactivates the pin. That is, the pin is active when LOW.

The PulseLowEnabledPin() function specifically does the following:

1. Calls the digitalWrite(PinNumber, LOW) function and writes a LOW value to the pin number PinNumber.

2. Calls the delayMicroseconds(DurationMicroSecs) function and halts the execution of the program for DurationMicroSecs microseconds.

3. Calls the digitalWrite(PinNumber, HIGH) function and writes a HIGH value to the pin number PinNumber.

4. Calls the delayMicroseconds(DurationMicroSecs) function and halts the execution of the program for DurationMicroSecs microseconds.

See Listing 5-14.

Listing 5-14. The PulseLowEnabledPin() function

```
void PulseLowEnabledPin(int PinNumber, int DurationMicroSecs)

{

// For Low Enabled Pins , 0 = on and 1 = off

digitalWrite(PinNumber, LOW);        // Sets the pin on

delayMicroseconds(DurationMicroSecs);   // Pauses for DurationMicroSecs microseconds


digitalWrite(PinNumber, HIGH);       // Sets the pin off
```

```
    delayMicroseconds(DurationMicroSecs);    // Pauses for DurationMicroSecs microseconds
}
```

The PulsePin() function creates a pulse by setting a pin HIGH for a period of time which activates the pin and then setting the pin back to a LOW value which deactivates the pin.

> Note: The PulsePin() function assumes a pin where a HIGH value activates the pin and a LOW value deactivates the pin. That is, the pin is active when HIGH.

Specifically the PulsePin() function does the following:

1. Calls the digitalWrite(PinNumber, HIGH) function that sets a pin to a HIGH value.

2. Calls the delayMicroseconds(DurationMicroSecs) function that halts the execution of the program for DurationMicroSecs microseconds.

3. Calls the digitalWrite(PinNumber, LOW) function that sets a pin to a LOW value.

4. Calls the delayMicroseconds(DurationMicroSecs) function that halts execution of the program for DurationMicroSecs microseconds.

See Listing 5-15.

Listing 5-15. The PulsePin() function

```
void PulsePin(int PinNumber, int DurationMicroSecs)
{
    digitalWrite(PinNumber, HIGH);           // Sets the pin on

    delayMicroseconds(DurationMicroSecs);    // Pauses for DurationMicroSecs microseconds


    digitalWrite(PinNumber, LOW);            // Sets the pin off

    delayMicroseconds(DurationMicroSecs);    // Pauses for DurationMicroSecs microseconds
}
```

SD Card Initialization

We are done now initializing the camera. The next thing we need to initialize is the SD card.

In order to initialize the SD card we need to:

1. Call the pinMode(HardwareSSPin, OUTPUT) function setting pin 53 which is the hardware slave select pin to OUTPUT. This must be done even though we are not using it because the SD library requires it.

2. Call the SD.begin(chipSelect) function with the chipSelect pin that can be used to enable or disable the SD card reader.

3. If the return value is false then the SD card was not successfully initialized. Otherwise, the card has been successfully initialized for use.

See Listing 5-16.

Listing 5-16. Initializing the SD card

```
// Initialize SD Card
Serial.print(F("\nInitializing SD card..."));
pinMode(HardwareSSPin, OUTPUT);    // change this to 53 on a mega


if (!SD.begin(chipSelect))
{
    Serial.println(F("Initialization failed ... /nThings to check:"));
    Serial.println(F("- Is a card is inserted?"));
    Serial.println(F("- Is your wiring correct?"));
    Serial.println(F("- Did you change the chipSelect pin to match your shield or module?"));
    return;
} else {
     Serial.println(F("Wiring is correct and a card is present ..."));
}
```

Main Command Loop

After the setup() function is called the loop() function is called repeatedly until the Arduino is reset or the Arduino is powered off.

The loop() function does the following:

1. Reads in user input from the Serial Monitor.

2. Execute various camera or SD card functions based on the user input

Reading in the User Input

The BUFFERLENGTH variable represents the length of the buffer that holds the characters to be read in.

The IncomingByte character array actually holds the user input data.

The RawCommandLine holds the String value of the user input data.

The user input loop is a while loop that does the following:

1. Calls the Serial.available() function and if there is user input to be read in which means the return value > 0 then read in the user input.

2. Reads in the user's input by calling the Serial.readBytesUntil('\n', IncomingByte, BUFFERLENGTH) function. The function reads in the user input until a newline character is found, BUFFERLENGTH number of characters have been read in, or the function times out. The function returns the user input characters in the IncomingByte character array.

3. Adds each character in the IncomingByte array to the RawCommandLine variable.

4. Breaks program execution out of the infinite while loop so that the user input can be processed.

See Listing 5-17.

Listing 5-17. Waiting for a user command.

```
// Serial Input

const int BUFFERLENGTH = 255;

char IncomingByte[BUFFERLENGTH];   // for incoming serial data

String RawCommandLine = "";



// Wait for Command

Serial.println(F("Ready to Accept new Command => "));

while (1)

{

    if (Serial.available() > 0)

    {

    int NumberCharsRead = Serial.readBytesUntil('\n', IncomingByte, BUFFERLENGTH);

    for (int i = 0; i < NumberCharsRead; i++)

    {

    RawCommandLine += IncomingByte[i];

    }

    break;
```

}

}

Processing User Input

Next, the user input is processed.

Processing the user input does the following:

1. Calls the DisplayHelpMenu() function if "h" or "help" is the user input.

2. Calls the DisplayHelpCommandsParams() function if "help camera" is the user input.

3. Calls the DisplayCurrentCommand() function if "d" is the user input.

4. Commands the camera to take a photo with the current camera command and parameters if "t" is the user input.

5. Calls the ReadPrintFile("TEST.TXT") function with "test.txt" as a filename if "testread" is the user input.

6. Writes a test file to the SD card if "testwrite" is the user input.

7. Changes the camera command and/or parameters if a camera command or parameter is the user input.

8. Resets the RawCommandLine which holds the user input to empty.

See Listing 5-18.

Listing 5-18. Processing User Input

```
// Print out the command from Android

Serial.print(F("Raw Command from Serial Monitor: "));

Serial.println(RawCommandLine);



if ((RawCommandLine == "h")||

   (RawCommandLine == "help"))

{

    DisplayHelpMenu();

}

else

if (RawCommandLine == "help camera")
```

```
{
    DisplayHelpCommandsParams();
}
else
if (RawCommandLine == "d")
{
    DisplayCurrentCommand();
}
else
if (RawCommandLine == "t")
{
    // Take Photo
    Serial.println(F("\nGoing to take photo with current command:"));
    DisplayCurrentCommand();

    // Take Photo
    ExecuteCommand(Command);
    Serial.println(F("Photo Taken and Saved to Arduino SD CARD ..."));

    String Testfile = CreatePhotoFilename();
    Serial.print(F("Image Output Filename :"));
    Serial.println(Testfile);
    PhotoTakenCount++;
}
else
if (RawCommandLine == "testread")
{
    ReadPrintFile("TEST.TXT");
}
```

```
else

if (RawCommandLine == "testwrite")

{

    CheckRemoveFile("TEST.TXT");

    WriteFileTest("TEST.TXT");

}
else
{

    Serial.println(F("Changing command or parameters according to your input:"));

    // Parse Command Line and Set Command Line Elements

    // Parse Raw Command into Command and Parameters

    ParseRawCommand(RawCommandLine);


    // Display new changed camera command with parameters

    DisplayCurrentCommand();

}


// Reset Command Line

RawCommandLine = "";


Serial.println();

Serial.println();
```

The Main Help Menu

The DisplayHelpMenu() function displays the main help menu for the program. The text is printing to the screen using the Serial.print() or Serial.println() functions. Also the F() indicates that the text is stored inside the Arduino's flash memory.

See Listing 5-19.

Listing 5-19. DisplayHelpMenu()

```
void  DisplayHelpMenu()
```

```
{
    Serial.println(F("................. Help Menu .................."));
    Serial.println(F("d - Display Current Camera Command"));
    Serial.println(F("t - Take Photograph using current Command and Parameters"));
    Serial.println(F("testread - Tests reading files from the SDCard by reading and printig the contents of test.txt"));
    Serial.println(F("testwrite - Tests writing files to SDCard"));
    Serial.println(F("help camera - Displays Camera's Commands and Parameters"));
    Serial.println();
    Serial.println();
}
```

Camera Help Menu

The DisplayHelpCommandsParams() function displays the help menu for the camera showing the set of commands that can change the camera's resolution as well as parameters for the frames per second settings, auto white balance settings, automatic exposure settings, YUV Matrix settings, denoise settings, edge enhancement setttings, and automatic black level calibration settings.

See Listing 5-20.

Listing 5-20. Camera Help Menu

```
void DisplayHelpCommandsParams()
{
    Serial.println(F("....... Help Menu Camera Commands/Params .........."));

    Serial.println(F("Resolution Change Commands: VGA,VGAP,QVGA,QQVGA"));
    Serial.println(F("FPS Parameters: ThirtyFPS, NightMode"));
    Serial.println(F("AWB Parameters: SAWB, AAWB"));
    Serial.println(F("AEC Parameters: AveAEC, HistAEC"));
    Serial.println(F("YUV Matrix Parameters: YUVMatrixOn, YUVMatrixOff"));
    Serial.println(F("Denoise Parameters: DenoiseYes, DenoiseNo"));
    Serial.println(F("Edge Enchancement: EdgeYes, EdgeNo"));
    Serial.println(F("Automatic Black Level Calibration: AblcON, AblcOFF"));
```

```
    Serial.println();

    Serial.println();

}
```

Displaying the Current Camera Command

The DisplayCurrentCommand() function displays the current settings for the camera command and camera parameters. The variables that hold the camera command and camera parameters are String variables and are shown in Listing 5-21 with their default values.

Listing 5-21. Camera command and parameter variables

```
String Command = "QQVGA";

String FPSParam = "ThirtyFPS";

String AWBParam = "SAWB";

String AECParam = "HistAEC";

String YUVMatrixParam = "YUVMatrixOn";

String DenoiseParam = "DenoiseNo";

String EdgeParam = "EdgeNo";

String ABLCParam = "AblcON";
```

The actual DisplayCurrentCommand() function is shown in Listing 5-22. The function consists of Serial.print() and Serial.println() statements that prints out the camera command and then the values of the camera parameters.

Listing 5-22. The DisplayCurrentCommand() function

```
void DisplayCurrentCommand()

{

    // Print out Command and Parameters

    Serial.println(F("Current Command:"));

    Serial.print(F("Command: "));

    Serial.println(Command);


    Serial.print(F("FPSParam: "));

    Serial.println(FPSParam);
```

```
    Serial.print(F("AWBParam: "));
    Serial.println(AWBParam);


    Serial.print(F("AECParam: "));
    Serial.println(AECParam);


    Serial.print(F("YUVMatrixParam: "));
    Serial.println(YUVMatrixParam);


    Serial.print(F("DenoiseParam: "));
    Serial.println(DenoiseParam);


    Serial.print(F("EdgeParam: "));
    Serial.println(EdgeParam);


    Serial.print(F("ABLCParam: "));
    Serial.println(ABLCParam);


    Serial.println();
}
```

Taking a Photo

If the user input is "t" then a photo is taken with the camera and saved to the SD card.

In order to take a photo with the camera the following steps are taken:

1. The DisplayCurrentCommand() function is called to display the current camera command and camera parameters in the Serial Monitor.

2. The ExecuteCommand(Command) function is called with the camera command as a parameter in order to take the actual photo.

3. The CreatePhotoFilename() function is called to retrieve the filename that the image was saved under and then prints this to the screen.

4. Increases the PhotoTakenCount variable that keeps track of the number of photos taken.

See Listing 5- 23.

Listing 5-23. Taking a photo

```
if (RawCommandLine == "t")
{
    // Take Photo
    Serial.println(F("\nGoing to take photo with current command:"));
    DisplayCurrentCommand();


    // Take Photo
    ExecuteCommand(Command);
    Serial.println(F("Photo Taken and Saved to Arduino SD CARD ..."));


    String Testfile = CreatePhotoFilename();
    Serial.print(F("Image Output Filename :"));
    Serial.println(Testfile);
    PhotoTakenCount++;
}
```

The ExecuteCommand() function uses the Resolution variable to check to see if the camera's current resolution settings need to be changed. If it does then a function is called to setup the camera according to the new resolution.

The current resolution of the camera is held in the Resolution variable that is of ResolutionType and by default is set to None.

The enumerated types are:

- VGA - VGA resolution and Raw Bayer format

- VGAP - VGA resolution and Raw Bayer format but the image is also processed through the Digital Signal Processor

- QVGA - QVGA resolution and YUV image format

- QQVGA - QQVGA resolution and YUV image format

- None - which means that the camera has not been set to any specific resolution yet

See Listing 5-24.

Listing 5-24. The Resolution variable

enum ResolutionType

{

VGA,

VGAP,

QVGA,

QQVGA,

None

};

ResolutionType Resolution = None;

The ExecuteCommand() function sets the camera to the correct resolution and parameters, then takes the photo and then saves it to an SD Card.

The ExecuteCommand() function does the following:

1. If the command is VGA then the camera takes a VGA photo. If the current resolution is not VGA or if parameters have changed then the camera registers are reset, the current resolution is set to VGA, the SetupOV7670ForVGARawRGB() function is called to actually set up the VGA mode, and the ReadRegisters() function is called to read key registers and print them out to screen.

2. If the command is VGAP then the camera takes a processed VGA photo. If the current resolution is not VGAP or if parameters have changed then the camera registers are reset, the current resolution is set to VGAP, the SetupOV7670ForVGAProcessedBayerRGB() function is called to actually set up the processed VGA mode, and the ReadRegisters() function is called to read key registers and print them out to screen.

3. If the command is QVGA then the camera takes a QVGA photo. If the current resolution is not QVGA or if parameters have changed then the current resolution is set to QVGA, the SetupOV7670ForQVGAYUV() function is called to actually set up the QVGA mode, and the ReadRegisters() function is called to read key registers and print them out to screen.

4. If the command is QQVGA then the camera takes a QQVGA photo. If the current resolution is not QQVGA or if parameters have changed then the current resolution is set to QQVGA, the SetupOV7670ForQQVGAYUV() function is called to actually set up the QQVGA mode, and the ReadRegisters() function is called to read key registers and print them out to screen.

5. The delay(100) function is called to halt the program execution for 100 milliseconds to allow time for the camera's registers to be completely updated after being set.

6. The TakePhoto() function is called that actually is responsible for taking the photo and saving it to an SD card.

See Listing 5-25.

Listing 5-25 The ExecuteCommand() function

```
void ExecuteCommand(String Command)

{

 // Set up Camera for VGA, QVGA, or QQVGA Modes

 if (Command == "VGA")

 {

   Serial.println(F("Taking a VGA Photo..."));

   if (Resolution != VGA)

   {

     // If current resolution is not QQVGA then set camera for QQVGA

     ResetCameraRegisters();


     Resolution = VGA;

     SetupOV7670ForVGARawRGB();

     Serial.println(F("----------------------------- Camera Registers ---------------------------"));

     ReadRegisters();

     Serial.println(F("-------------------------------------------------------------------------"));

   }

 }

 else

 if (Command == "VGAP")

 {

   Serial.println(F("Taking a VGAP Photo..."));

   if (Resolution != VGAP)

   {

     // If current resolution is not VGAP then set camera for VGAP

     ResetCameraRegisters();


     Resolution = VGAP;
```

```
      SetupOV7670ForVGAProcessedBayerRGB();
      Serial.println(F("--------------------------- Camera Registers ---------------------------"));
      ReadRegisters();
      Serial.println(F("--------------------------------------------------------------------"));
    }
  }
  else if (Command == "QVGA")
  {
    Serial.println(F("Taking a QVGA Photo..."));
    if (Resolution != QVGA)
    {
      // If current resolution is not QQVGA then set camera for QQVGA
      Resolution = QVGA;
      SetupOV7670ForQVGAYUV();
      Serial.println(F("--------------------------- Camera Registers ---------------------------"));
      ReadRegisters();
      Serial.println(F("--------------------------------------------------------------------"));
    }
  }
  else if (Command == "QQVGA")
  {
    Serial.println(F("Taking a QQVGA Photo..."));
    if (Resolution != QQVGA)
    {
      // If current resolution is not QQVGA then set camera for QQVGA
      Resolution = QQVGA;
      SetupOV7670ForQQVGAYUV();
      Serial.println(F("--------------------------- Camera Registers ---------------------------"));
      ReadRegisters();
```

```
      Serial.println(F("--------------------------------------------------------------------"));

    }

  }

  else

  {

    Serial.print(F("The command "));

    Serial.print(Command);

    Serial.println(F(" is not recognized ..."));

  }


  // Delay for registers to settle

  delay(100);


  // Take Photo

  TakePhoto();

}
```

The SetupOV7670ForVGARawRGB() function sets the ov7670 camera for the output of a raw RGB Bayer image in VGA resolution.

The SetupOV7670ForVGARawRGB() function specifically does the following:

1. Sets the PHOTO_WIDTH variable that represents the width of the output image to 640. This variable is later used to read in the image data from the camera's memory.

2. Sets the PHOTO_HEIGHT variable that represents the height of the output image to 480. This variable is later used to read in the image data from the camera's memory.

3. Sets the PHOTO_BYTES_PER_PIXEL that represents the bytes per pixel of the output image to 1. This variable is later used to read in the image data from the camera's memory.

4. The width, height, and bytes per pixel of the output image are then printed to the screen on the Serial Monitor.

5. The VGA screen resolution mode is then activated by setting values for the camera registers by calling the OV7670WriteReg() function for each register needed with the associated new value and then ParseI2CResult(result) is called with the result of the write operation. This returns a text error status message that is then printed out. The registers that need to be set are:

    1. CLKRC

2. COM7
3. COM3
4. COM14
5. SCALING_XSC
6. SCALING_YSC
7. SCALING_DCWCTR
8. SCALING_PCLK_DIV
9. SCALING_PCLK_DELAY
10. COM17

6. Calls the SetCameraFPSMode() function to determine if the camera frame rate will be 30 frames per second or a Night Mode where the frame rate varies with the amount of light available.

7. Calls the SetCameraAEC() function that sets the type of automatic exposure control that is either average based or histogram based.

8. Calls the OV7670WriteReg(0xB0, 0x8c) function and sets a register that is listed in the official documentation as unused or "reserved" which is 0xB0 to the value of 0x8C. This is done to correct the colors that should be red but appear as green without this register adjustment.

9. Calls the SetCameraSaturationControl() function that sets the saturation control for the camera.

10. Calls the SetupCameraArrayControl() function that sets the camera's array control registers.

11. Calls the SetupCameraADCControl() function that sets the camera's analog to digital converter related registers.

12. Calls the SetupCameraABLC() function that sets the automatic black level correction.

13. Calls the OV7670WriteReg() function and sets the window output registers which are the following to values for the VGA screen resolution.

    1. HSTART
    2. HSTOP
    3. HREF
    4. VSTRT
    5. VSTOP
    6. VREF

See Listing 5-26.

Listing 5-26. The SetupOV7670ForVGARawRGB() function

```
void SetupOV7670ForVGARawRGB()

{

  int result = 0;

  String sresult = "";


  Serial.println(F("-------------------------- Setting Camera for VGA (Raw RGB) --------------------------"));


  PHOTO_WIDTH  = 640;

  PHOTO_HEIGHT = 480;

  PHOTO_BYTES_PER_PIXEL = 1;


  Serial.print(F("Photo Width = "));

  Serial.println(PHOTO_WIDTH);


  Serial.print(F("Photo Height = "));

  Serial.println(PHOTO_HEIGHT);


  Serial.print(F("Bytes Per Pixel = "));

  Serial.println(PHOTO_BYTES_PER_PIXEL);


  // Basic Registers

  result = OV7670WriteReg(CLKRC, CLKRC_VALUE_VGA);

  sresult = ParseI2CResult(result);

  Serial.print(F("CLKRC: "));

  Serial.println(sresult);


  result = OV7670WriteReg(COM7, COM7_VALUE_VGA );
```

```
//result = OV7670WriteReg(COM7, COM7_VALUE_VGA_COLOR_BAR );
sresult = ParseI2CResult(result);
Serial.print(F("COM7: "));
Serial.println(sresult);


result = OV7670WriteReg(COM3, COM3_VALUE_VGA);
sresult = ParseI2CResult(result);
Serial.print(F("COM3: "));
Serial.println(sresult);


result = OV7670WriteReg(COM14, COM14_VALUE_VGA );
sresult = ParseI2CResult(result);
Serial.print(F("COM14: "));
Serial.println(sresult);


result = OV7670WriteReg(SCALING_XSC,SCALING_XSC_VALUE_VGA );
sresult = ParseI2CResult(result);
Serial.print(F("SCALING_XSC: "));
Serial.println(sresult);


result = OV7670WriteReg(SCALING_YSC,SCALING_YSC_VALUE_VGA );
sresult = ParseI2CResult(result);
Serial.print(F("SCALING_YSC: "));
Serial.println(sresult);


result = OV7670WriteReg(SCALING_DCWCTR, SCALING_DCWCTR_VALUE_VGA );
sresult = ParseI2CResult(result);
Serial.print(F("SCALING_DCWCTR: "));
Serial.println(sresult);
```

```
result = OV7670WriteReg(SCALING_PCLK_DIV, SCALING_PCLK_DIV_VALUE_VGA);

sresult = ParseI2CResult(result);

Serial.print(F("SCALING_PCLK_DIV: "));

Serial.println (sresult);


result = OV7670WriteReg(SCALING_PCLK_DELAY,SCALING_PCLK_DELAY_VALUE_VGA);

sresult = ParseI2CResult(result);

Serial.print(F("SCALING_PCLK_DELAY: "));

Serial.println(sresult);


// COM17 - DSP Color Bar Enable/Disable

// COM17_VALUE  0x08 // Activate Color Bar for DSP

//result = OV7670WriteReg(COM17, COM17_VALUE_AEC_NORMAL_COLOR_BAR);

result = OV7670WriteReg(COM17, COM17_VALUE_AEC_NORMAL_NO_COLOR_BAR);

sresult = ParseI2CResult(result);

Serial.print(F("COM17: "));

Serial.println(sresult);


// Set Additional Parameters

// Set Camera Frames per second

SetCameraFPSMode();


// Set Camera Automatic Exposure Control

SetCameraAEC();


// Needed Color Correction, green to red

result = OV7670WriteReg(0xB0, 0x8c);

sresult = ParseI2CResult(result);
```

```
Serial.print(F("Setting B0 UNDOCUMENTED register to 0x84:= "));
Serial.println(sresult);


// Set Camera Saturation
SetCameraSaturationControl();


// Setup Camera Array Control
SetupCameraArrayControl();


// Set ADC Control
SetupCameraADCControl();


// Set Automatic Black Level Calibration
SetupCameraABLC();


Serial.println(F(".......... Setting Camera Window Output Parameters ........"));
// Change Window Output parameters after custom scaling
result = OV7670WriteReg(HSTART, HSTART_VALUE_VGA );
sresult = ParseI2CResult(result);
Serial.print(F("HSTART: "));
Serial.println(sresult);


result = OV7670WriteReg(HSTOP, HSTOP_VALUE_VGA );
sresult = ParseI2CResult(result);
Serial.print(F("HSTOP: "));
Serial.println(sresult);


result = OV7670WriteReg(HREF, HREF_VALUE_VGA );
sresult = ParseI2CResult(result);
```

```
    Serial.print(F("HREF: "));

    Serial.println(sresult);


    result = OV7670WriteReg(VSTRT, VSTRT_VALUE_VGA );

    sresult = ParseI2CResult(result);

    Serial.print(F("VSTRT: "));

    Serial.println(sresult);


    result = OV7670WriteReg(VSTOP, VSTOP_VALUE_VGA );

    sresult = ParseI2CResult(result);

    Serial.print(F("VSTOP: "));

    Serial.println(sresult);


    result = OV7670WriteReg(VREF, VREF_VALUE_VGA );

    sresult = ParseI2CResult(result);

    Serial.print(F("VREF: "));

    Serial.println(sresult);

}
```

The SetCameraFPSMode() function sets the number of FPS (frames per second that the camera captures) for the camera by:

1. Calling the SetupCameraFor30FPS() function if the FPSParam variable is set to "ThirtyFPS".

2. Calling the SetupCameraNightMode() function if the FPSParam variable is set to "NightMode".

See Listing 5-27.

Listing 5-27. The SetCameraFPSMode() function

```
void SetCameraFPSMode()

{

    // Set FPS for Camera

    if (FPSParam == "ThirtyFPS")
```

```
    {
      SetupCameraFor30FPS();
    }
    else
    if (FPSParam == "NightMode")
    {
      SetupCameraNightMode();
    }
}
```

The SetupCameraFor30FPS() function sets the camera to capture images continuously at the rate of 30 frames per second.

The registers that are set are:

1. CLKRC
2. DBLV
3. EXHCH
4. EXHCL
5. DM_LNL
6. DM_LNH
7. COM11

See Listing 5-28.

Listing 5-28. SetupCameraFor30FPS

```
void SetupCameraFor30FPS()
{
  int result = 0;
  String sresult = "";


  Serial.println(F("........... Setting Camera to 30 FPS ........"));
  result = OV7670WriteReg(CLKRC, CLKRC_VALUE_30FPS);
  sresult = ParseI2CResult(result);
```

```
Serial.print(F("CLKRC: "));

Serial.println(sresult);


result = OV7670WriteReg(DBLV, DBLV_VALUE_30FPS);

sresult = ParseI2CResult(result);

Serial.print(F("DBLV: "));

Serial.println(sresult);


result = OV7670WriteReg(EXHCH, EXHCH_VALUE_30FPS);

sresult = ParseI2CResult(result);

Serial.print(F("EXHCH: "));

Serial.println(sresult);


result = OV7670WriteReg(EXHCL, EXHCL_VALUE_30FPS);

sresult = ParseI2CResult(result);

Serial.print(F("EXHCL: "));

Serial.println(sresult);


result = OV7670WriteReg(DM_LNL, DM_LNL_VALUE_30FPS);

sresult = ParseI2CResult(result);

Serial.print(F("DM_LNL: "));

Serial.println(sresult);


result = OV7670WriteReg(DM_LNH, DM_LNH_VALUE_30FPS);

sresult = ParseI2CResult(result);

Serial.print(F("DM_LNH: "));

Serial.println(sresult);


result = OV7670WriteReg(COM11, COM11_VALUE_30FPS);
```

```
  sresult = ParseI2CResult(result);

  Serial.print(F("COM11: "));

  Serial.println(sresult);
}
```

The SetupCameraNightMode() function sets the camera to night mode where the frame speed is adjusted automatically to match the amount of light in the environment.

The camera registers set are:

1. CLKRC
2. COM11

See Listing 5-29.

Listing 5-29. The SetupCameraNightMode() function

```
void SetupCameraNightMode()
{
  int result = 0;
  String sresult = "";

  Serial.println(F("......... Turning NIGHT MODE ON ........"));
  result = OV7670WriteReg(CLKRC, CLKRC_VALUE_NIGHTMODE_AUTO);
  sresult = ParseI2CResult(result);
  Serial.print(F("CLKRC: "));
  Serial.println(sresult);


  result = OV7670WriteReg(COM11, COM11_VALUE_NIGHTMODE_AUTO);
  sresult = ParseI2CResult(result);
  Serial.print(F("COM11: "));
  Serial.println(sresult);
}
```

The SetCameraAEC() function sets the automatic exposure control method to either the average method or the histogram method based on the value of the AECParam.

The function calls either:

1. SetupCameraAverageBasedAECAGC() if the average method is selected by the user or

2. SetCameraHistogramBasedAECAGC() if the histogram method is selected by the user.

See Listing 5-30.

Listing 5-30. The SetCameraAEC() function

```
void SetCameraAEC()
{
  // Process AEC
  if (AECParam == "AveAEC")
  {
    // Set Camera's Average AEC/AGC Parameters
    SetupCameraAverageBasedAECAGC();
  }
  else
  if (AECParam == "HistAEC")
  {
    // Set Camera AEC algorithim to Histogram
    SetCameraHistogramBasedAECAGC();
  }
}
```

The SetupCameraAverageBasedAECAGC() function sets the automatic exposure control for the average method.

The camera registers affected are:

1. AEW

2. AEB

3. VPT

4. HAECC7

All these registers are set and the result is returned and printed out. See Listing 5-31.

Listing 5-31. The SetupCameraAverageBasedAECAGC() function

```
void SetupCameraAverageBasedAECAGC()
{
  int result = 0;
  String sresult = "";

  Serial.println(F("-------------- Setting Camera Average Based AEC/AGC Registers --------------"));

  result = OV7670WriteReg(AEW, AEW_VALUE);
  sresult = ParseI2CResult(result);
  Serial.print(F("AEW: "));
  Serial.println(sresult);

  result = OV7670WriteReg(AEB, AEB_VALUE);
  sresult = ParseI2CResult(result);
  Serial.print(F("AEB: "));
  Serial.println(sresult);

  result = OV7670WriteReg(VPT, VPT_VALUE);
  sresult = ParseI2CResult(result);
  Serial.print(F("VPT: "));
  Serial.println(sresult);

  result = OV7670WriteReg(HAECC7, HAECC7_VALUE_AVERAGE_AEC_ON);
  sresult = ParseI2CResult(result);
  Serial.print(F("HAECC7: "));
  Serial.println(sresult);
}
```

The SetCameraHistogramBasedAECAGC() function sets the automatic exposure control to the histogram method.

The registers affected are:

1. AEW
2. AEB
3. HAECC1
4. HAECC2
5. HAECC3
6. HAECC4
7. HAECC5
8. HAECC6
9. HAECC7

All these registers are set and the result is returned and print out. See Listing 5-32.

Listing 5-32. The SetCameraHistogramBasedAECAGC() function

```
void SetCameraHistogramBasedAECAGC()
{
  int result = 0;
  String sresult = "";

  Serial.println(F("-------------- Setting Camera Histogram Based AEC/AGC Registers ---------------"));

  result = OV7670WriteReg(AEW, AEW_VALUE);
  sresult = ParseI2CResult(result);
  Serial.print(F("AEW: "));
  Serial.println(sresult);


  result = OV7670WriteReg(AEB, AEB_VALUE);
  sresult = ParseI2CResult(result);
  Serial.print(F("AEB: "));
```

```
Serial.println(sresult);

result = OV7670WriteReg(HAECC1, HAECC1_VALUE);
sresult = ParseI2CResult(result);
Serial.print(F("HAECC1: "));
Serial.println(sresult);

result = OV7670WriteReg(HAECC2, HAECC2_VALUE);
sresult = ParseI2CResult(result);
Serial.print(F("HAECC2: "));
Serial.println(sresult);

result = OV7670WriteReg(HAECC3, HAECC3_VALUE);
sresult = ParseI2CResult(result);
Serial.print(F("HAECC3: "));
Serial.println(sresult);

result = OV7670WriteReg(HAECC4, HAECC4_VALUE);
sresult = ParseI2CResult(result);
Serial.print(F("HAECC4: "));
Serial.println(sresult);

result = OV7670WriteReg(HAECC5, HAECC5_VALUE);
sresult = ParseI2CResult(result);
Serial.print(F("HAECC5: "));
Serial.println(sresult);

result = OV7670WriteReg(HAECC6, HAECC6_VALUE);
sresult = ParseI2CResult(result);
```

```
    Serial.print(F("HAECC6: "));

    Serial.println(sresult);


    result = OV7670WriteReg(HAECC7, HAECC7_VALUE_HISTOGRAM_AEC_ON);

    sresult = ParseI2CResult(result);

    Serial.print(F("HAECC7: "));

    Serial.println(sresult);

}
```

The SetCameraSaturationControl() function controls the level of saturation in the image by setting the saturation control register SATCTR. See Listing 5-33.

Listing 5-33. The SetCameraSaturationControl() function

```
void SetCameraSaturationControl()
{
    int result = 0;
    String sresult = "";


    Serial.println(F(".......... Setting Camera Saturation Level ........"));

    result = OV7670WriteReg(SATCTR, SATCTR_VALUE);

    sresult = ParseI2CResult(result);

    Serial.print(F("SATCTR: "));

    Serial.println(sresult);

}
```

The SetupCameraArrayControl() function sets the camera array registers:

1. CHLF and
2. ARBLM

See Listing 5-34.

Listing 5-34. The SetupCameraArrayControl() function

```
void SetupCameraArrayControl()
```

```
{
    int result = 0;

    String sresult = "";


    Serial.println(F("........... Setting Camera Array Control ........"));


    result = OV7670WriteReg(CHLF, CHLF_VALUE);

    sresult = ParseI2CResult(result);

    Serial.print(F("CHLF: "));

    Serial.println(sresult);


    result = OV7670WriteReg(ARBLM, ARBLM_VALUE);

    sresult = ParseI2CResult(result);

    Serial.print(F("ARBLM: "));

    Serial.println(sresult);

}
```

The SetupCameraADCControl() function sets up the camera's analog to digital conversion controls relating to the conversion of the image from analog to digital.

The SetupCameraADCControl() function affects camera registers:

1. ADCCTR1

2. ADCCTR2

3. ADC

4. ACOM

5. OFON

The registers are written to using the OV7670WriteReg() function, the return error status is processed using the ParseI2CResult() function and the results are printed out to the screen. See Listing 5-35.

Listing 5-35. The SetupCameraADCControl() function

```
void SetupCameraADCControl()
```

```
{
    int result = 0;
    String sresult = "";

    Serial.println(F("........... Setting Camera ADC Control  ........"));

    result = OV7670WriteReg(ADCCTR1, ADCCTR1_VALUE);
    sresult = ParseI2CResult(result);
    Serial.print(F("ADCCTR1: "));
    Serial.println(sresult);

    result = OV7670WriteReg(ADCCTR2, ADCCTR2_VALUE);
    sresult = ParseI2CResult(result);
    Serial.print(F("ADCCTR2: "));
    Serial.println(sresult);

    result = OV7670WriteReg(ADC, ADC_VALUE);
    sresult = ParseI2CResult(result);
    Serial.print(F("ADC: "));
    Serial.println(sresult);

    result = OV7670WriteReg(ACOM, ACOM_VALUE);
    sresult = ParseI2CResult(result);
    Serial.print(F("ACOM: "));
    Serial.println(sresult);

    result = OV7670WriteReg(OFON, OFON_VALUE);
    sresult = ParseI2CResult(result);
    Serial.print(F("OFON: "));
```

Serial.println(sresult);

}

The SetupCameraABLC() function sets the automatic black level calibration based on the value of ABLCParam.

The camera registers affected are:

1. ABLC1

2. THL_ST

The registers are written to using the OV7670WriteReg() function, the error status is sent to ParseI2CResult() , and the returned string form of the error status is printed out. See Listing 5-36.

Listing 5-36. The SetupCameraABLC() function

```
void SetupCameraABLC()

{

  int result = 0;

  String sresult = "";


  // If ABLC is off then return otherwise
  // turn on ABLC.
  if (ABLCParam == "AblcOFF")
  {
    return;
  }


  Serial.println(F("........ Setting Camera ABLC ........"));


  result = OV7670WriteReg(ABLC1, ABLC1_VALUE);
  sresult = ParseI2CResult(result);
  Serial.print(F("ABLC1: "));
  Serial.println(sresult);
```

```
    result = OV7670WriteReg(THL_ST, THL_ST_VALUE);

    sresult = ParseI2CResult(result);

    Serial.print(F("THL_ST: "));

    Serial.println(sresult);

}
```

The SetupOV7670ForVGAProcessedBayerRGB() function sets the camera for VGA resolution and the camera's output to a processed Bayer RGB image.

The SetupOV7670ForVGAProcessedBayerRGB() function does the following:

1. The SetupOV7670ForVGARawRGB() function is called to set up the camera for VGA resolution.

2. The register COM7 is set with the value for activating processed Bayer RGB as an output image.

3. The register TSLB is set with the value needed to correct incorrectly displayed colors.

4. The register 0xB0 is set with 0x8c in order to correct incorrectly displayed colors.

5. The SetupCameraAWB() function is called to set the automatic white balance according to the user's parameters.

6. The SetupCameraDenoiseEdgeEnhancement() function is called to set the denoise and edge enhancement according to the user's parameters.

See Listing 5-37.

Listing 5-37. The SetupOV7670ForVGAProcessedBayerRGB() function

```
void SetupOV7670ForVGAProcessedBayerRGB()

{

  int result = 0;

  String sresult = "";



  // Call Base for VGA Raw Bayer RGB Mode

  SetupOV7670ForVGARawRGB();



  Serial.println(F("------------- Setting Camera for VGA (Processed Bayer RGB) ----------------"));



  // Set key register for selecting processed bayer rgb output
```

```
    result = OV7670WriteReg(COM7, COM7_VALUE_VGA_PROCESSED_BAYER );

    //result = OV7670WriteReg(COM7, COM7_VALUE_VGA_COLOR_BAR );

    sresult = ParseI2CResult(result);

    Serial.print(F("COM7: "));

    Serial.println(sresult);


    result = OV7670WriteReg(TSLB, 0x04);

    sresult = ParseI2CResult(result);

    Serial.print(F("Initializing TSLB register result = "));

    Serial.println(sresult);


    // Needed Color Correction, green to red

    result = OV7670WriteReg(0xB0, 0x8c);

    sresult = ParseI2CResult(result);

    Serial.print(F("Setting B0 UNDOCUMENTED register to 0x84:= "));

    Serial.println(sresult);


    // Set Camera Automatic White Balance

    SetupCameraAWB();


    // Denoise and Edge Enhancement

    SetupCameraDenoiseEdgeEnhancement();
}
```

The SetupCameraAWB() function sets the type of automatic white balance to either simple or advanced depending on the AWBParam variable.

The SetupCameraAWB() function does the following:

1. If the AWBParam is "SAWB" then the SetupCameraSimpleAutomaticWhiteBalance() function is called followed by the SetupCameraGain() function.

2. If the AWBParam is "AAWB" then the SetupCameraAdvancedAutomaticWhiteBalance() function is called followed by the SetupCameraAdvancedAutoWhiteBalanceConfig() and then the SetupCameraGain() function is called.

See Listing 5-38.

Listing 5-38. The SetupCameraAWB() function

```
void SetupCameraAWB()

{

  // Set AWB Mode

  if (AWBParam == "SAWB")

  {

    // Set Simple Automatic White Balance

    SetupCameraSimpleAutomaticWhiteBalance(); // OK



    // Set Gain Config

    SetupCameraGain();

  }

  else

  if (AWBParam == "AAWB")

  {

    // Set Advanced Automatic White Balance

    SetupCameraAdvancedAutomaticWhiteBalance(); // ok



    // Set Camera Automatic White Balance Configuration

    SetupCameraAdvancedAutoWhiteBalanceConfig(); // ok



    // Set Gain Config

    SetupCameraGain();

  }

}
```

The SetupCameraSimpleAutomaticWhiteBalance() function sets the automatic white balance type to simple by:

1. Setting the COM8 register using the OV7670WriteReg() function and then printing out the result of the write operation after it is converted into text using the ParseI2Cresult() function.

2. Setting the AWBCTR0 register using the OV7670WriteReg() function and then printing out the result of the write operation after it is converted into text using the ParseI2Cresult() function.

See Listing 5-39.

Listing 5-39. The SetupCameraSimpleAutomaticWhiteBalance() function

```
void SetupCameraSimpleAutomaticWhiteBalance()

{

  int result = 0;

  String sresult = "";


  Serial.println(F("........... Setting Camera to Simple AWB ........"));


  // COM8

  result = OV7670WriteReg(COM8, COM8_VALUE_AWB_ON);

  sresult = ParseI2CResult(result);

  Serial.print(F("COM8(0x13): "));

  Serial.println(sresult);


  // AWBCTR0

  result = OV7670WriteReg(AWBCTR0, AWBCTR0_VALUE_NORMAL);

  sresult = ParseI2CResult(result);

  Serial.print(F("AWBCTR0 Control Register 0(0x6F): "));

  Serial.println(sresult);

}
```

The SetupCameraGain() function sets the gain for the camera by:

1. Setting the COM9 register using the OV7670WriteReg() function and printing out the success or failure status of the write operation.

2. Setting the BLUE register using the OV7670WriteReg() function and printing out the success or failure status of the write operation.

3. Setting the RED register using the OV7670WriteReg() function and printing out the success or failure status of the write operation.

4. Setting the GGAIN register using the OV7670WriteReg() function and printing out the success or failure status of the write operation.

5. Setting the COM16 register using the OV7670WriteReg() function and printing out the success or failure status of the write operation.

See Listing 5-40.

Listing 5-40. The SetupCameraGain() function

```
void SetupCameraGain()

{

  int result = 0;

  String sresult = "";


  Serial.println(F("........... Setting Camera Gain ........"));


  // Set Maximum Gain

  result = OV7670WriteReg(COM9, COM9_VALUE_4XGAIN);

  sresult = ParseI2CResult(result);

  Serial.print(F("COM9: "));

  Serial.println(sresult);


  // Set Blue Gain

  result = OV7670WriteReg(BLUE, BLUE_VALUE);

  sresult = ParseI2CResult(result);

  Serial.print(F("BLUE GAIN: "));

  Serial.println(sresult);


  // Set Red Gain
```

```
    result = OV7670WriteReg(RED, RED_VALUE);

    sresult = ParseI2CResult(result);

    Serial.print(F("RED GAIN: "));

    Serial.println(sresult);


    // Set Green Gain

    result = OV7670WriteReg(GGAIN, GGAIN_VALUE);

    sresult = ParseI2CResult(result);

    Serial.print(F("GREEN GAIN: "));

    Serial.println(sresult);


    // Enable AWB Gain

    result = OV7670WriteReg(COM16, COM16_VALUE);

    sresult = ParseI2CResult(result);

    Serial.print(F("COM16(ENABLE GAIN): "));

    Serial.println(sresult);

}
```

The SetupCameraAdvancedAutomaticWhiteBalance() function sets the automatic white balance setting to type Advanced by:

1. Setting the COM8 register using the OV7670WriteReg() function and printing out the success or failure status of the write operation.

2. Setting the AWBCTR0 register using the OV7670WriteReg() function and printing out the success or failure status of the write operation.

See Listing 5-41.

Listing 5-41. The SetupCameraAdvancedAutomaticWhiteBalance() function

```
void SetupCameraAdvancedAutomaticWhiteBalance()
{
    int result = 0;

    String sresult = "";
```

```
Serial.println(F("........... Setting Camera to Advanced AWB ........"));

// AGC, AWB, and AEC Enable

result = OV7670WriteReg(0x13, 0xE7);

sresult = ParseI2CResult(result);

Serial.print(F("COM8(0x13): "));

Serial.println(sresult);


// AWBCTR0

result = OV7670WriteReg(0x6f, 0x9E);

sresult = ParseI2CResult(result);

Serial.print(F("AWB Control Register 0(0x6F): "));

Serial.println(sresult);

}
```

The SetupCameraAdvancedAutoWhiteBalanceConfig() function sets some more registers that are required for the advanced automatic white balance by setting the following registers and displaying the success or failure status of each write operation:

1. AWBC1
2. AWBC2
3. AWBC3
4. AWBC4
5. AWBC5
6. AWBC6
7. AWBC7
8. AWBC8
9. AWBC9
10. AWBC10
11. AWBC11
12. AWBC12

13. AWBCTR3

14. AWBCTR2

15. AWBCTR1

See Listing 5-42.

Listing 5-42. The SetupCameraAdvancedAutoWhiteBalanceConfig() function

```
void SetupCameraAdvancedAutoWhiteBalanceConfig()
{
  int result = 0;
  String sresult = "";

  Serial.println(F("........... Setting Camera Advanced Auto White Balance Configs ........"));

  result = OV7670WriteReg(AWBC1, AWBC1_VALUE);
  sresult = ParseI2CResult(result);
  Serial.print(F("AWBC1: "));
  Serial.println(sresult);

  result = OV7670WriteReg(AWBC2, AWBC2_VALUE);
  sresult = ParseI2CResult(result);
  Serial.print(F("AWBC2: "));
  Serial.println(sresult);

  result = OV7670WriteReg(AWBC3, AWBC3_VALUE);
  sresult = ParseI2CResult(result);
  Serial.print(F("AWBC3: "));
  Serial.println(sresult);

  result = OV7670WriteReg(AWBC4, AWBC4_VALUE);
  sresult = ParseI2CResult(result);
```

```
Serial.print(F("AWBC4: "));

Serial.println(sresult);


result = OV7670WriteReg(AWBC5, AWBC5_VALUE);

sresult = ParseI2CResult(result);

Serial.print(F("AWBC5: "));

Serial.println(sresult);


result = OV7670WriteReg(AWBC6, AWBC6_VALUE);

sresult = ParseI2CResult(result);

Serial.print(F("AWBC6: "));

Serial.println(sresult);


result = OV7670WriteReg(AWBC7, AWBC7_VALUE);

sresult = ParseI2CResult(result);

Serial.print(F("AWBC7: "));

Serial.println(sresult);


result = OV7670WriteReg(AWBC8, AWBC8_VALUE);

sresult = ParseI2CResult(result);

Serial.print(F("AWBC8: "));

Serial.println(sresult);


result = OV7670WriteReg(AWBC9, AWBC9_VALUE);

sresult = ParseI2CResult(result);

Serial.print(F("AWBC9: "));

Serial.println(sresult);


result = OV7670WriteReg(AWBC10, AWBC10_VALUE);
```

```cpp
    sresult = ParseI2CResult(result);
    Serial.print(F("AWBC10: "));
    Serial.println(sresult);


    result = OV7670WriteReg(AWBC11, AWBC11_VALUE);
    sresult = ParseI2CResult(result);
    Serial.print(F("AWBC11: "));
    Serial.println(sresult);


    result = OV7670WriteReg(AWBC12, AWBC12_VALUE);
    sresult = ParseI2CResult(result);
    Serial.print(F("AWBC12: "));
    Serial.println(sresult);


    result = OV7670WriteReg(AWBCTR3, AWBCTR3_VALUE);
    sresult = ParseI2CResult(result);
    Serial.print(F("AWBCTR3: "));
    Serial.println(sresult);


    result = OV7670WriteReg(AWBCTR2, AWBCTR2_VALUE);
    sresult = ParseI2CResult(result);
    Serial.print(F("AWBCTR2: "));
    Serial.println(sresult);


    result = OV7670WriteReg(AWBCTR1, AWBCTR1_VALUE);
    sresult = ParseI2CResult(result);
    Serial.print(F("AWBCTR1: "));
    Serial.println(sresult);
}
```

The SetupCameraDenoiseEdgeEnhancement() function performs denoising and edge enhancement on an image depending on the values of the DenoiseParam and EdgeParam variables.

The SetupCameraDenoiseEdgeEnhancement() function does the following:

1. If the DenoiseParam is "DenoiseYes" and the EdgeParam is "EdgeYes" then both the denoise and edge enhancement image options have been selected by the user. The SetupCameraDenoise() function is called to set the camera registers that will denoise the image. The SetupCameraEdgeEnhancement() function is called to set to turn on the edge enhancement function of the camera. The COM16 register is set through the OV7670WriteReg() function with the value that will turn both the denoise and edge enhancement features on. Specifically the value is in the #define COM16_VALUE_DENOISE_ON__EDGE_ENHANCEMENT_ON__AWBGAIN_ON.

2. If the DenoiseParam is "DenoiseYes" and the EdgeParam is "EdgeNo" then the user has selected denoising but has turned off edge enhancement. The SetupCameraDenoise() function is then called to turn on the camera's denoising function. The COM16 register is set through the OV7670WriteReg() function. The value to be written to the COM16 register is from the #define COM16_VALUE_DENOISE_ON__EDGE_ENHANCEMENT_OFF__AWBGAIN_ON.

3. If the DenoiseParam is "DenoiseNo" and the EdgeParam is "EdgeYes" then the user has selected to turn denoising off and to turn on edge enhancement. Then the SetupCameraEdgeEnhancement() function is called that turns on the edge enhancement. The COM16 register is set with the function OV7670WriteReg() with the value that specified by the #define COM16_VALUE_DENOISE_OFF__EDGE_ENHANCEMENT_ON__AWBGAIN_ON

    Important Note: the COM16 contains the control bits for both the denoise and edge enhance operations. Bit 5 controls the edge enhancement and bit 4 controls the denoise operation.

See Listing 5-42.

Listing 5-42. The SetupCameraDenoiseEdgeEnhancement(0 function

```
void SetupCameraDenoiseEdgeEnhancement()

{

  int result = 0;

  String sresult = "";


  if ((DenoiseParam == "DenoiseYes")&&

    (EdgeParam == "EdgeYes"))

    {
```

```
    SetupCameraDenoise();
    SetupCameraEdgeEnhancement();
    result = OV7670WriteReg(COM16,
COM16_VALUE_DENOISE_ON__EDGE_ENHANCEMENT_ON__AWBGAIN_ON);
    sresult = ParseI2CResult(result);
    Serial.print(F("COM16: "));
    Serial.println(sresult);
  }
  else
  if ((DenoiseParam == "DenoiseYes")&&
     (EdgeParam == "EdgeNo"))
  {
    SetupCameraDenoise();
    result = OV7670WriteReg(COM16,
COM16_VALUE_DENOISE_ON__EDGE_ENHANCEMENT_OFF__AWBGAIN_ON);
    sresult = ParseI2CResult(result);
    Serial.print(F("COM16: "));
    Serial.println(sresult);
  }
  else
  if ((DenoiseParam == "DenoiseNo")&&
     (EdgeParam == "EdgeYes"))
   {
     SetupCameraEdgeEnhancement();
     result = OV7670WriteReg(COM16,
COM16_VALUE_DENOISE_OFF__EDGE_ENHANCEMENT_ON__AWBGAIN_ON);
     sresult = ParseI2CResult(result);
     Serial.print(F("COM16: "));
     Serial.println(sresult);
    }
}
```

The SetupCameraDenoise() function sets the camera's registers so that the denoise function is activated for image processing.

The function sets the following camera registers:

1. DNSTH

2. REG77

See Listing 5-43.

Listing 5-43. The SetupCameraDenoise() function

```
void SetupCameraDenoise()
{
  int result = 0;
  String sresult = "";


  Serial.println(F("........... Setting Camera Denoise ........"));


  result = OV7670WriteReg(DNSTH, DNSTH_VALUE);
  sresult = ParseI2CResult(result);
  Serial.print(F("DNSTH: "));
  Serial.println(sresult);


  result = OV7670WriteReg(REG77, REG77_VALUE);
  sresult = ParseI2CResult(result);
  Serial.print(F("REG77: "));
  Serial.println(sresult);
}
```

The SetupCameraEdgeEnhancement() function activates the camera's image edge enhancement by:

1. Setting the EDGE camera register.

2. Setting the REG75 camera register.

3. Setting the REG76 camera register.

See Listing 5-44.

Listing 5-44. The SetupCameraEdgeEnhancement() function

```
void SetupCameraEdgeEnhancement()
{
  int result = 0;

  String sresult = "";


    Serial.println(F("........... Setting Camera Edge Enhancement ........"));


    result = OV7670WriteReg(EDGE, EDGE_VALUE);

    sresult = ParseI2CResult(result);

    Serial.print(F("EDGE: "));

    Serial.println(sresult);


    result = OV7670WriteReg(REG75, REG75_VALUE);

    sresult = ParseI2CResult(result);

    Serial.print(F("REG75: "));

    Serial.println(sresult);


    result = OV7670WriteReg(REG76, REG76_VALUE);

    sresult = ParseI2CResult(result);

    Serial.print(F("REG76: "));

    Serial.println(sresult);
}
```

The SetupOV7670ForQVGAYUV() function sets up the camera to take pictures in QVGA resolution and output the image in YUV format.

The SetupOV7670ForQVGAYUV() does the following:

1. The PHOTO_WIDTH variable which represents the width in pixels of the captured camera image is set to 320. This variable is used in reading in the captured image from the camera's FIFO memory.

2. The PHOTO_HEIGHT variable which represents the height in pixels of the captured camera image is set to 240. This variable is used in reading in the captured image from the camera's FIFO memory.

3. The PHOTO_BYTES_PER_PIXEL variable which represents the number of bytes per pixel of the captured image is set to 2. This variable is used in reading in the captured image from the camera's FIFO memory.

4. The CLKRC camera register is set.

5. The COM7 camera register is set.

6. The COM3 camera register is set.

7. The COM14 camera register is set.

8. The SCALING_XSC camera register is set.

9. The SCALING_YSC camera register is set.

10. The SCALING_DCWCTR camera register is set.

11. The SCALING_PCLK_DIV camera register is set.

12. The SCALING_PCLK_DELAY camera register is set.

13. The TSLB camera register is set.

14. The COM13 camera register is set.

15. The COM17 camera register is set.

16. The SetCameraFPSMode() function sets the frames per second for the camera based on user selections.

17. The SetCameraAEC() function sets the automatic exposure control type of the camera based on user selections.

18. The SetupCameraAWB() function sets the automatic white balance controls of the camera based on user selections.

19. The SetupCameraUndocumentedRegisters() function sets the undocumented registers that are needed for the camera to operate correctly.

20. If the YUVMatrixParam is "YUVMatrixOn" then the SetCameraColorMatrixYUV() function that sets the color matrix specifically for a YUV image is called.

21. The SetCameraSaturationControl() function sets the camera's saturation control.

22. The SetupCameraDenoiseEdgeEnhancement() function sets the camera's denoise and edge enhancement image processing functions.

23. The SetupCameraArrayControl() function sets the camera's array controls.

24. The SetupCameraADCControl() function sets the camera's analog to digital conversion controls.

25. The SetupCameraABLC() function sets the camera's automatic black level calibration settings.

26. The camera's output window registers are set to QVGA values. The registers are:

    1. HSTART
    2. HSTOP
    3. HREF
    4. VSTRT
    5. VSTOP
    6. VREF

See Listing 5-45.

Listing 5-45. The SetupOV7670ForQVGAYUV() function

```
void SetupOV7670ForQVGAYUV()

{

  int result = 0;

  String sresult = "";


  Serial.println(F("--------------------------- Setting Camera for QVGA (YUV) ---------------------------"));


  PHOTO_WIDTH  = 320;

  PHOTO_HEIGHT = 240;

  PHOTO_BYTES_PER_PIXEL = 2;


  Serial.print(F("Photo Width = "));

  Serial.println(PHOTO_WIDTH);


  Serial.print(F("Photo Height = "));

  Serial.println(PHOTO_HEIGHT);
```

```
Serial.print(F("Bytes Per Pixel = "));

Serial.println(PHOTO_BYTES_PER_PIXEL);



// Basic Registers

result = OV7670WriteReg(CLKRC, CLKRC_VALUE_QVGA);

sresult = ParseI2CResult(result);

Serial.print(F("CLKRC: "));

Serial.println(sresult);


result = OV7670WriteReg(COM7, COM7_VALUE_QVGA );

//result = OV7670WriteReg(COM7, COM7_VALUE_QVGA_COLOR_BAR );

sresult = ParseI2CResult(result);

Serial.print(F("COM7: "));

Serial.println(sresult);


result = OV7670WriteReg(COM3, COM3_VALUE_QVGA);

sresult = ParseI2CResult(result);

Serial.print(F("COM3: "));

Serial.println(sresult);


result = OV7670WriteReg(COM14, COM14_VALUE_QVGA );

sresult = ParseI2CResult(result);

Serial.print(F("COM14: "));

Serial.println(sresult);


result = OV7670WriteReg(SCALING_XSC,SCALING_XSC_VALUE_QVGA );

sresult = ParseI2CResult(result);
```

```
Serial.print(F("SCALING_XSC: "));
Serial.println(sresult);


result = OV7670WriteReg(SCALING_YSC,SCALING_YSC_VALUE_QVGA );
sresult = ParseI2CResult(result);
Serial.print(F("SCALING_YSC: "));
Serial.println(sresult);


result = OV7670WriteReg(SCALING_DCWCTR, SCALING_DCWCTR_VALUE_QVGA );
sresult = ParseI2CResult(result);
Serial.print(F("SCALING_DCWCTR: "));
Serial.println(sresult);


result = OV7670WriteReg(SCALING_PCLK_DIV, SCALING_PCLK_DIV_VALUE_QVGA);
sresult = ParseI2CResult(result);
Serial.print(F("SCALING_PCLK_DIV: "));
Serial.println (sresult);


result = OV7670WriteReg(SCALING_PCLK_DELAY,SCALING_PCLK_DELAY_VALUE_QVGA );
sresult = ParseI2CResult(result);
Serial.print(F("SCALING_PCLK_DELAY: "));
Serial.println(sresult);


// YUV order control change from default use with COM13
result = OV7670WriteReg(TSLB, 0x04);
sresult = ParseI2CResult(result);
Serial.print(F("TSLB: "));
Serial.println(sresult);
```

//COM13

result = OV7670WriteReg(COM13, 0xC2);   // from YCbCr reference specs

sresult = ParseI2CResult(result);

Serial.print(F("COM13: "));

Serial.println(sresult);

// COM17 - DSP Color Bar Enable/Disable

// COM17_VALUE 0x08 // Activate Color Bar for DSP

//result = OV7670WriteReg(COM17, COM17_VALUE_AEC_NORMAL_COLOR_BAR);

result = OV7670WriteReg(COM17, COM17_VALUE_AEC_NORMAL_NO_COLOR_BAR);

sresult = ParseI2CResult(result);

Serial.print(F("COM17: "));

Serial.println(sresult);

// Set Additional Parameters

// Set Camera Frames per second

SetCameraFPSMode();

// Set Camera Automatic Exposure Control

SetCameraAEC();

// Set Camera Automatic White Balance

SetupCameraAWB();

// Setup Undocumented Registers - Needed Minimum

SetupCameraUndocumentedRegisters();

// Set Color Matrix for YUV

if (YUVMatrixParam == "YUVMatrixOn")

```
{
  SetCameraColorMatrixYUV();
}

// Set Camera Saturation
SetCameraSaturationControl();

// Denoise and Edge Enhancement
SetupCameraDenoiseEdgeEnhancement();

// Set up Camera Array Control
SetupCameraArrayControl();

// Set ADC Control
SetupCameraADCControl();

// Set Automatic Black Level Calibration
SetupCameraABLC();

Serial.println(F("........... Setting Camera Window Output Parameters ........"));
// Change Window Output parameters after custom scaling
result = OV7670WriteReg(HSTART, HSTART_VALUE_QVGA );
sresult = ParseI2CResult(result);
Serial.print(F("HSTART: "));
Serial.println(sresult);


result = OV7670WriteReg(HSTOP, HSTOP_VALUE_QVGA );
sresult = ParseI2CResult(result);
Serial.print(F("HSTOP: "));
```

```
    Serial.println(sresult);


    result = OV7670WriteReg(HREF, HREF_VALUE_QVGA );

    sresult = ParseI2CResult(result);

    Serial.print(F("HREF: "));

    Serial.println(sresult);


    result = OV7670WriteReg(VSTRT, VSTRT_VALUE_QVGA );

    sresult = ParseI2CResult(result);

    Serial.print(F("VSTRT: "));

    Serial.println(sresult);


    result = OV7670WriteReg(VSTOP, VSTOP_VALUE_QVGA );

    sresult = ParseI2CResult(result);

    Serial.print(F("VSTOP: "));

    Serial.println(sresult);


    result = OV7670WriteReg(VREF, VREF_VALUE_QVGA );

    sresult = ParseI2CResult(result);

    Serial.print(F("VREF: "));

    Serial.println(sresult);

}
```

The SetupCameraUndocumentedRegisters() function sets the register 0xB0 with the value 0x84. This is an undocumented register that was listed as "reserved" in the main documentation. However, the "OV7670 Software Application Note" document specified this value under the "Reference Settings" section under "YcbCr". See Listing 5-46.

Listing 5-46. The SetupCameraUndocumentedRegisters() function

```
void SetupCameraUndocumentedRegisters()

{

    int result = 0;
```

```
    String sresult = "";


    Serial.println(F("........... Setting Camera Undocumented Registers ........"));

    result = OV7670WriteReg(0xB0, 0x84);

    sresult = ParseI2CResult(result);

    Serial.print(F("Setting B0 UNDOCUMENTED register to 0x84:= "));

    Serial.println(sresult);
}
```

The SetCameraColorMatrixYUV() function sets the color matrix for the YUV output image by setting camera registers:

1. MTX1 though MTX6
2. CONTRAS
3. MTXS

See Listing 5-47.

Listing 5-47. The SetCameraColorMatrixYUV() function

```
void SetCameraColorMatrixYUV()
{
    int result = 0;

    String sresult = "";


    Serial.println(F("........... Setting Camera Color Matrix for YUV ........"));


    result = OV7670WriteReg(MTX1, MTX1_VALUE);

    sresult = ParseI2CResult(result);

    Serial.print(F("MTX1: "));

    Serial.println(sresult);


    result = OV7670WriteReg(MTX2, MTX2_VALUE);

    sresult = ParseI2CResult(result);
```

```
Serial.print(F("MTX2: "));
Serial.println(sresult);


result = OV7670WriteReg(MTX3, MTX3_VALUE);
sresult = ParseI2CResult(result);
Serial.print(F("MTX3: "));
Serial.println(sresult);


result = OV7670WriteReg(MTX4, MTX4_VALUE);
sresult = ParseI2CResult(result);
Serial.print(F("MTX4: "));
Serial.println(sresult);


result = OV7670WriteReg(MTX5, MTX5_VALUE);
sresult = ParseI2CResult(result);
Serial.print(F("MTX5: "));
Serial.println(sresult);


result = OV7670WriteReg(MTX6, MTX6_VALUE);
sresult = ParseI2CResult(result);
Serial.print(F("MTX6: "));
Serial.println(sresult);


result = OV7670WriteReg(CONTRAS, CONTRAS_VALUE);
sresult = ParseI2CResult(result);
Serial.print(F("CONTRAS: "));
Serial.println(sresult);


result = OV7670WriteReg(MTXS, MTXS_VALUE);
```

```
    sresult = ParseI2CResult(result);

    Serial.print(F("MTXS: "));

    Serial.println(sresult);

}
```

The SetupOV7670ForQQVGAYUV() function sets the camera to take photos in QQVGA resolution in the YUV format. More specifically the YUV image is in yuyv order with 2 bytes per pixel in the image.

The SetupOV7670ForQQVGAYUV() function does the following:

1. Sets the PHOTO_WIDTH variable which represents the captured photo's width to 160. This variable is used to read in the image from the camera's frame buffer FIFO memory.

2. Sets the PHOTO_HEIGHT variable which represents the captured photo's height to 120. This variable is used to read in the image from the camera's frame buffer FIFO memory.

3. Set the PHOTO_BYTES_PER_PIXEL variable which represents the captured photo's bytes per pixel to 2. This variable is used to read in the image from the camera's frame buffer FIFO memory.

4. Sets the camera's CLKRC register.

5. Sets the camera's COM7 register.

6. Sets the camera's COM3 register.

7. Sets the camera's COM14 register.

8. Sets the camera's SCALING_XSC register.

9. Sets the camera's SCALING_YSC register.

10. Sets the camera's SCALING_DCWCTR register.

11. Sets the camera's SCALING_PCLK_DIV register.

12. Sets the camera's SCALING_PCLK_DELAY register.

13. Sets the camera's TSLB register.

14. Sets the camera's COM13 register.

15. Sets the camera's COM17 register.

16. The SetCameraFPSMode() function is called and sets the frames per second mode of the camera.

17. The SetCameraAEC() function is called and sets the automatic exposure controls for the camera.

18. The SetupCameraAWB() function is called and sets the camera's automatic white balance controls for the camera.

19. The SetupCameraUndocumentedRegisters() function is called and sets the camera's undocumented registers that are needed to produce a correct color photo output.

20. if the YUVMatrixParam variable is "YUVMatrixOn" then the SetCameraColorMatrixYUV() function is called to set the camera's color matrix for YUV output.

21. The SetCameraSaturationControl() function is called to set the saturation control for the camera.

22. The SetupCameraDenoiseEdgeEnhancement() function is called to set the denoise and edge enhancement settings based on user selections.

23. The SetupCameraArrayControl() function is called to set the camera's array control.

24. The SetupCameraADCControl() function is called to set the camera's analog to digital conversion controls.

25. The SetupCameraABLC() function is called to set the automatic black level calibration of the camera.

26. The output window dimensions are set to QQVGA by modifying registers:

    1. HSTART
    2. HSTOP
    3. HREF
    4. VSTRT
    5. VSTOP
    6. VREF

See Listing 5-48.

Listing 5-48. The SetupOV7670ForQQVGAYUV() function

```
void SetupOV7670ForQQVGAYUV()

{
    int result = 0;

    String sresult = "";



    Serial.println(F("-------------------------- Setting Camera for QQVGA YUV --------------------------"));
```

```
PHOTO_WIDTH  = 160;
PHOTO_HEIGHT = 120;
PHOTO_BYTES_PER_PIXEL = 2;


Serial.print(F("Photo Width = "));
Serial.println(PHOTO_WIDTH);


Serial.print(F("Photo Height = "));
Serial.println(PHOTO_HEIGHT);


Serial.print(F("Bytes Per Pixel = "));
Serial.println(PHOTO_BYTES_PER_PIXEL);



Serial.println(F("........... Setting Basic QQVGA Parameters ........"));


result = OV7670WriteReg(CLKRC, CLKRC_VALUE_QQVGA);
sresult = ParseI2CResult(result);
Serial.print(F("CLKRC: "));
Serial.println(sresult);


result = OV7670WriteReg(COM7, COM7_VALUE_QQVGA );
//result = OV7670WriteReg(COM7, COM7_VALUE_QQVGA_COLOR_BAR );
sresult = ParseI2CResult(result);
Serial.print(F("COM7: "));
Serial.println(sresult);


result = OV7670WriteReg(COM3, COM3_VALUE_QQVGA);
sresult = ParseI2CResult(result);
```

```
    Serial.print(F("COM3: "));
    Serial.println(sresult);


    result = OV7670WriteReg(COM14, COM14_VALUE_QQVGA );
    sresult = ParseI2CResult(result);
    Serial.print(F("COM14: "));
    Serial.println(sresult);


    result = OV7670WriteReg(SCALING_XSC,SCALING_XSC_VALUE_QQVGA );
    sresult = ParseI2CResult(result);
    Serial.print(F("SCALING_XSC: "));
    Serial.println(sresult);


    result = OV7670WriteReg(SCALING_YSC,SCALING_YSC_VALUE_QQVGA );
    sresult = ParseI2CResult(result);
    Serial.print(F("SCALING_YSC: "));
    Serial.println(sresult);


    result = OV7670WriteReg(SCALING_DCWCTR, SCALING_DCWCTR_VALUE_QQVGA );
    sresult = ParseI2CResult(result);
    Serial.print(F("SCALING_DCWCTR: "));
    Serial.println(sresult);


    result = OV7670WriteReg(SCALING_PCLK_DIV, SCALING_PCLK_DIV_VALUE_QQVGA);
    sresult = ParseI2CResult(result);
    Serial.print(F("SCALING_PCLK_DIV: "));
    Serial.println (sresult);


    result = OV7670WriteReg(SCALING_PCLK_DELAY,SCALING_PCLK_DELAY_VALUE_QQVGA );
```

```
sresult = ParseI2CResult(result);
Serial.print(F("SCALING_PCLK_DELAY: "));
Serial.println(sresult);


// YUV order control change from default use with COM13
result = OV7670WriteReg(TSLB, TSLB_VALUE_YUYV_AUTO_OUTPUT_WINDOW_DISABLED); // Works
sresult = ParseI2CResult(result);
Serial.print(F("TSLB: "));
Serial.println(sresult);


//COM13
result = OV7670WriteReg(COM13, 0xC8);  // Gamma Enabled, UV Auto Adj On
sresult = ParseI2CResult(result);
Serial.print(F("COM13: "));
Serial.println(sresult);


// COM17 - DSP Color Bar Enable/Disable
//result = OV7670WriteReg(COM17, COM17_VALUE_AEC_NORMAL_COLOR_BAR);
result = OV7670WriteReg(COM17, COM17_VALUE_AEC_NORMAL_NO_COLOR_BAR);
sresult = ParseI2CResult(result);
Serial.print(F("COM17: "));
Serial.println(sresult);


// Set Additional Parameters
// Set Camera Frames per second
SetCameraFPSMode();


// Set Camera Automatic Exposure Control
SetCameraAEC();
```

```
// Set Camera Automatic White Balance
SetupCameraAWB();


// Setup Undocumented Registers - Needed Minimum
SetupCameraUndocumentedRegisters();


// Set Color Matrix for YUV
if (YUVMatrixParam == "YUVMatrixOn")
{
  SetCameraColorMatrixYUV();
}


// Set Camera Saturation
SetCameraSaturationControl();


// Denoise and Edge Enhancement
SetupCameraDenoiseEdgeEnhancement();


// Set Array Control
SetupCameraArrayControl();


// Set ADC Control
SetupCameraADCControl();


// Set Automatic Black Level Calibration
SetupCameraABLC();


Serial.println(F(".......... Setting Camera Window Output Parameters ........"));
```

```
// Change Window Output parameters after custom scaling
result = OV7670WriteReg(HSTART, HSTART_VALUE_QQVGA );
sresult = ParseI2CResult(result);
Serial.print(F("HSTART: "));
Serial.println(sresult);


result = OV7670WriteReg(HSTOP, HSTOP_VALUE_QQVGA );
sresult = ParseI2CResult(result);
Serial.print(F("HSTOP: "));
Serial.println(sresult);


result = OV7670WriteReg(HREF, HREF_VALUE_QQVGA );
sresult = ParseI2CResult(result);
Serial.print(F("HREF: "));
Serial.println(sresult);


result = OV7670WriteReg(VSTRT, VSTRT_VALUE_QQVGA );
sresult = ParseI2CResult(result);
Serial.print(F("VSTRT: "));
Serial.println(sresult);


result = OV7670WriteReg(VSTOP, VSTOP_VALUE_QQVGA );
sresult = ParseI2CResult(result);
Serial.print(F("VSTOP: "));
Serial.println(sresult);


result = OV7670WriteReg(VREF, VREF_VALUE_QQVGA );
sresult = ParseI2CResult(result);
Serial.print(F("VREF: "));
```

```
  Serial.println(sresult);

}
```

The TakePhoto() function takes the photo with the camera and then saves it to the SD card.

The TakePhoto() function does the following:

1. Records the start time in milliseconds by calling the millis() function which returns the number of milliseconds since the current program has been started.

2. The CaptureOV7670Frame() function is called to take a photo with the ov7670 camera which is saved in the camera's FIFO frame buffer memory.

3. The ReadTransmitCapturedFrame() function is called to save the image in the camera's memory to an SD card.

4. The EndTime variable records the end time by calling the millis() function.

5. The ElapsedTime variable is set to the time it takes for a photo to be captured to the camera's memory and written to the SD card.

See Listing 5-49.

Listing 5-49. The TakePhoto() function

```
void TakePhoto()

{

 unsigned long StartTime   = 0;

 unsigned long EndTime     = 0;

 unsigned long ElapsedTime = 0;


 StartTime = millis();


 CaptureOV7670Frame();

 ReadTransmitCapturedFrame();


 EndTime = millis();

 ElapsedTime = (EndTime - StartTime)/1000; // Convert to seconds


 Serial.print(F("Elapsed Time for Taking and Sending Photo(secs) = "));
```

Serial.println(ElapsedTime);

}

The CaptureOV7670Frame() function captures an image from the ov7670 camera and saves it to the camera's FIFO frame buffer memory.

The CaptureOV7670FrameI() function does the following:

1. Waits for the VSync input from the camera to pulse to indicate the start of the image by calling the pulseIn(VSYNC, HIGH) function. The pulseIn() function makes the Arduino wait for the pin VSYNC to pulse from LOW to HIGH and then to LOW again before returning and continuing program execution.

2. Resets the FIFO memory frame buffer's write pointer to 0 that represents the beginning of the frame. The PulseLowEnabledPin(WRST, 6) function resets the write pointer by sending a LOW pulse to pin WRST for 6 microseconds.

3. Sets the FIFO write enable to active (high) so that image can be written to ram. This is done by calling the digitalWrite(WEN, HIGH) function that sets the output pin WEN to HIGH.

4. Waits for VSync to pulse again to indicate the end of the frame capture. The pulseIn(VSYNC, HIGH) function is called and waits for a positive pulse on the VSYNC pin.

5. Sets the FIFO write enable to nonactive (low) so that no more images can be written to the camera's memory. The digitalWrite(WEN, LOW) function is called to write a LOW value to pin WEN which is connected to the camera's write enable pin WR.

6. Prints out the elapsed time from the start of the image capture to the end of the image capture.

7. Program execution is halted for 2 milliseconds so that new data can appear on output pins. The delay() function is used to do this.

See Listing 5-50.

Listing 5-50. The CaptureOV7670Frame() function

```
void CaptureOV7670Frame()

{

  unsigned long DurationStart = 0;

  unsigned long DurationStop = 0;

  unsigned long TimeForCaptureStart = 0;

  unsigned long TimeForCaptureEnd = 0;

  unsigned long ElapsedTime = 0;



  //Capture one frame into FIFO memory
```

```
// 0. Initialization.
Serial.println();
Serial.println(F("Starting Capture of Photo ..."));
TimeForCaptureStart = millis();


// 1. Wait for VSync to pulse to indicate the start of the image
DurationStart = pulseIn(VSYNC, HIGH);


// 2. Reset Write Pointer to 0. Which is the beginning of frame
PulseLowEnabledPin(WRST, 6); // 3 microseconds + 3 microseconds for error factor on Arduino


// 3. Set FIFO Write Enable to active (high) so that image can be written to ram
digitalWrite(WEN, HIGH);


// 4. Wait for VSync to pulse again to indicate the end of the frame capture
DurationStop = pulseIn(VSYNC, HIGH);


// 5. Set FIFO Write Enable to nonactive (low) so that no more images can be written to the ram
digitalWrite(WEN, LOW);


// 6. Print out Stats
TimeForCaptureEnd = millis();
ElapsedTime = TimeForCaptureEnd - TimeForCaptureStart;

Serial.print(F("Time for Frame Capture (milliseconds) = "));
Serial.println(ElapsedTime);


Serial.print(F("VSync beginning duration (microseconds) = "));
Serial.println(DurationStart);
```

```
Serial.print(F("VSync end duration (microseconds) = "));

Serial.println(DurationStop);


// 7. WAIT so that new data can appear on output pins Read new data.

delay(2);

}
```

The ReadTransmitCapturedFrame() function reads in the image from the camera's memory and saves the image to an SD card.

The ReadTransmitCapturedFrame() function does the following:

1. The CreatePhotoFilename() function is called to create the name of the photo to save on the SD card based on certain parameters such as resolution, and photo number.

2. Calls the CheckRemoveFile(Filename) function to check if the file already exists and removes it if it does.

3. The SD.open(Filename.c_str(), FILE_WRITE) function opens a file on the SD card for writing and returns the opened file if successful.

4. If the returned file is null then exit the function.

5. Sets the read buffer pointer to the start of the camera's image frame. This done by calling the digitalWrite(RRST, LOW) function to reset the read pointer by setting the low active RRST pin to LOW. The read clock is pulsed in order to execute the read pointer reset. The read clock is pulsed using the PulsePin(RCLK, 1) function where the RCLK pin is pulsed for 1 microsecond. This is done three times. The RRST pin is then deactivated by setting the pin to HIGH. This is done by calling the digitalWrite(RRST, HIGH) function.

6. Pulses the read clock RCLK to bring in a new byte of data by calling the PulsePin(RCLK, 1) function.

7. Converts the input pin values from the camera's video port to byte values for pins 0-7 of the incoming pixel's byte. The ConvertPinValueToByteValue(digitalRead(DO7), 7) function is used to do this with the first parameter being the value of an input bit and the second parameter is the position of this bit.

8. Combines the individual pieces of data from each pin into composite data in the form of a single byte called PixelData.

9. Each byte from the image is saved using the write(PixelData) command that is called from the file object that was returned from Step 3. The parameter is the byte from the image data that came out of the camera's FIFO memory from Step 6.

10. Steps 6 through 9 are repeated until all the bytes from the image are read in and saved to the SD card.

11. The image file on the SD card is closed by calling the close() function from the file object that was opened in step 3. For example, ImageOutputFile.close().

12. Writes the photo's info file that contains the exact camera settings in a text file to the SDCard by calling the CreatePhotoInfoFile() function.

See Listing 5-51.

Listing 5-51. The ReadTransmitCapturedFrame() function

```
void ReadTransmitCapturedFrame()
{
  byte PixelData = 0;
  byte PinVal7 = 0;
  byte PinVal6 = 0;
  byte PinVal5 = 0;
  byte PinVal4 = 0;
  byte PinVal3 = 0;
  byte PinVal2 = 0;
  byte PinVal1 = 0;
  byte PinVal0 = 0;


  Serial.println(F("Starting Transmission of Photo To SDCard ..."));


  ///////////////////// Code for SD Card /////////////////////////////
  // Image file to write to
  File ImageOutputFile;


  // Create name of Photo To save based on certain parameters
  String Filename = CreatePhotoFilename();


  // Check if file already exists and remove it if it does.
  CheckRemoveFile(Filename);

  ImageOutputFile = SD.open(Filename.c_str(), FILE_WRITE);
```

```
// Test if file actually open
if (!ImageOutputFile)
{
  Serial.println(F("\nCritical ERROR ... Can not open Image Ouput File for output ... "));
   return;
}
///////////////////////////////////////////////////////////////


// Set Read Buffer Pointer to start of frame
digitalWrite(RRST, LOW);
PulsePin(RCLK, 1);
PulsePin(RCLK, 1);
PulsePin(RCLK, 1);
digitalWrite(RRST, HIGH);


unsigned long  ByteCounter = 0;
for (int height = 0; height < PHOTO_HEIGHT; height++)
{
  for (int width = 0; width < PHOTO_WIDTH; width++)
  {
    for (int bytenumber = 0; bytenumber < PHOTO_BYTES_PER_PIXEL; bytenumber++)
    {
      // Pulse the read clock RCLK to bring in new byte of data.
      PulsePin(RCLK, 1);

      // Convert Pin values to byte values for pins 0-7 of incoming pixel byte
      PinVal7 = ConvertPinValueToByteValue(digitalRead(DO7), 7);
      PinVal6 = ConvertPinValueToByteValue(digitalRead(DO6), 6);
```

```
        PinVal5 = ConvertPinValueToByteValue(digitalRead(DO5), 5);

        PinVal4 = ConvertPinValueToByteValue(digitalRead(DO4), 4);

        PinVal3 = ConvertPinValueToByteValue(digitalRead(DO3), 3);

        PinVal2 = ConvertPinValueToByteValue(digitalRead(DO2), 2);

        PinVal1 = ConvertPinValueToByteValue(digitalRead(DO1), 1);

        PinVal0 = ConvertPinValueToByteValue(digitalRead(DO0), 0);


        // Combine individual data from each pin into composite data in the form of a single byte

        PixelData = PinVal7 | PinVal6 | PinVal5 | PinVal4 | PinVal3 | PinVal2 | PinVal1 | PinVal0;


        ////////////////////////////  SD Card ////////////////////////////

        ByteCounter = ByteCounter + ImageOutputFile.write(PixelData);

        ///////////////////////////////////////////////////////////////////
      }
    }
  }
  // Close SD Card File
  ImageOutputFile.close();


  Serial.print(F("Total Bytes Saved to SDCard = "));
  Serial.println(ByteCounter);


  // Write Photo's Info File to SDCard.
  Serial.println(F("Writing Photo's Info file (.txt file) to SD Card ..."));
  CreatePhotoInfoFile();
}
```

The CreatePhotoFilename() function creates and returns a string that contains the file name for the photo that is to saved to the SD card.

The CreatePhotoFilename() function does the following:

1. If the camera command is "QQVGA" or the camera command is "QVGA" then the file extension is ".yuv" which stands for a YUV file.
2. If the camera command is "VGA" or the command is "VGAP" then the file extension is ".raw" which stands for a Bayer RAW file
3. The final file name for the photo is a concatenation of the camera command and the number of photos taken and the extension.

See Listing 5-52.

Listing 5-52 The CreatePhotoFilename() function

String CreatePhotoFilename()

{

  String Filename = "";

  String Ext = "";

  // Creates filename that the photo will be saved under

  // Create file extension

  // If Command = QQVGA or QVGA then extension is .yuv

  if ((Command == "QQVGA") || (Command == "QVGA"))

  {

    Ext = ".yuv";

  }

  else

  if ((Command == "VGA") || (Command == "VGAP"))

  {

    Ext = ".raw";

  }

  // Create Filename from

  // Resolution + PhotoNumber + Extension

  Filename = Command + PhotoTakenCount + Ext;

return Filename;

}

The CheckRemoveFile() function checks to see if a file exists and then removes it if it does.

The CheckRemoveFile() function does the following:

1. If the file exists on the SD card then remove it. That is if the SD.exists(tempchar) function returns true then call the SD.remove(tempchar) function with the file name of the file to remove from the SD card.
2. If the file still exists after trying to remove it then the new image file cannot be saved to the SD Card. An error message is printed out and the function is exited without trying to save the photo to the SD card.

See Listing 5-53.

Listing 5-53. The CheckRemoveFile() function

```
void CheckRemoveFile(String Filename)
{
  // Check if file already exists and remove it if it does.
  char tempchar[50];
  strcpy(tempchar, Filename.c_str());


  if (SD.exists(tempchar))
  {
    Serial.print(F("Filename: "));
    Serial.print(tempchar);
    Serial.println(F(" Already Exists. Removing It..."));
    SD.remove(tempchar);
  }


  // If file still exists then new image file cannot be saved to SD Card.
  if (SD.exists(tempchar))
  {
```

```
      Serial.println(F("Error.. Image output file cannot be created..."));

    return;

   }

 }
```

The ConvertPinValueToByteValue() function converts pin HIGH/LOW values on pins at positions 0-7 to a corresponding byte value. If the input PinValue is HIGH then the return value is a byte with the value of a 1 shifted left PinPosition number positions. If the PinValue is LOW then a zero value is returned. See Listing 5-54.

Listing 5-54. The ConvertPinValueToByteValue() function

```
byte ConvertPinValueToByteValue(int PinValue, int PinPosition)

{

  byte ByteValue = 0;

  if (PinValue == HIGH)

  {

    ByteValue = 1 << PinPosition;

  }

  return ByteValue;

}
```

The CreatePhotoInfoFile() function creates the photo information file based on current camera settings.

The CreatePhotoInfoFile() function does the following:

1. The CreatePhotoInfoFilename() function is called and returns a string with the name of the photo's information filename.

2. Checks if the file already exists and removes it if it does by calling the CheckRemoveFile(Filename) function with the filename of the file to check as a parameter.

3. Opens a writable file with the filename of the info file returned from Step1. The function SD.open(Filename.c_str(), FILE_WRITE) is used to do this.

4. Test if file has been opened successfully. If the file has not been opened successfully then exit the function.

5. Create the string that represents the camera setting information to be included in the photo's info file by calling CreatePhotoInfo() function.

6. Write the data returned from Step 5 to the info file opened in Step 3. The InfoFile.println(Data) function with the camera data being sent as the parameter is used to do this.

7. Close the SD Card File using the InfoFile.close() function

See Listing 5-55.

Listing 5-55. The CreatePhotoInfoFile() function

void CreatePhotoInfoFile()

{

  // Creates the photo information file based on current settings

  // .txt information File for Photo

  File InfoFile;

  // Create name of Photo To save based on certain parameters

  String Filename = CreatePhotoInfoFilename();

  // Check if file already exists and remove it if it does.

  CheckRemoveFile(Filename);

  // Open File

  InfoFile = SD.open(Filename.c_str(), FILE_WRITE);

  // Test if file actually open

  if (!InfoFile)

  {

    Serial.println(F("\nCritical ERROR ... Can not open Photo Info File for output ... "));

    return;

  }

  // Write Info to file

  String Data = CreatePhotoInfo();

  InfoFile.println(Data);

```
// Close SD Card File

  InfoFile.close();

}
```

The CreatePhotoInfoFilename() function creates the filename that the information about the photo will be saved under.

The CreatePhotoInfoFilename() function does the following:

1. Sets the file extension for an Info file for a photo to ".txt".

2. Create the final filename from the combination of the camera resolution, photo number, and Extension.

See Listing 5-56.

Listing 5-56. The CreatePhotoInfoFilename() function

```
String CreatePhotoInfoFilename()

{

  String Filename = "";

  String Ext = "";


  // Creates filename that the information about the photo will

  // be saved under.

  // Create file extension

  Ext = ".txt";


  // Create Filename from

  // Resolution + PhotoNumber + Extension

  Filename = Command + PhotoTakenCount + Ext;


  return Filename;

}
```

The CreatePhotoInfo() function creates the information which will be saved in a photo's info file that consists of the camera's command and the camera's parameters. See Listing 5-57.

Listing 5-57. The CreatePhotoInfo() function

```
String CreatePhotoInfo()
{
  String Info = "";

  Info = Command + " " + FPSParam + " " + AWBParam + " " + AECParam + " " + YUVMatrixParam + " " +
    DenoiseParam + " " + EdgeParam + " " + ABLCParam;

  return Info;
}
```

Testing the SD Card: Reading From the SD Card

The ReadPrintFile() function is used to test the reading capabilities of the SD card.

The ReadPrintFile() function does the following:

1. A file is opened on the SD card by calling the SD.open(Filename.c_str()) function.

2. If the file object that is returned is valid then read and print out all the file contents to the screen. This is done in a while loop that continues looping until there is no more information to read, that is the TempFile.available() function returns 0. The file data is written to the screen of the Serial Monitor by calling the Serial.write(TempFile.read()) function with the input parameter being file data that is read from the read() function on the File object.

3. The file is then closed by calling the TempFile.close() function on the File object that was opened in Step 1.

See Listing 5-58.

Listing 5-58. The ReadPrintFile() function

```
void ReadPrintFile(String Filename)
{
  File TempFile;

  // Reads in file and prints it to screen via Serial
  TempFile = SD.open(Filename.c_str());

  if (TempFile)
```

```
  {
    Serial.print(Filename);
    Serial.println(":");

    // read from the file until there's nothing else in it:
    while (TempFile.available())
    {
      Serial.write(TempFile.read());
    }
    // close the file:
    TempFile.close();
  }
  else
  {
    // Error opening file
    Serial.print("Error opening ");
    Serial.println(Filename);
  }
}
```

Testing the SD Card: Writing a File to the SD Card

The WriteFileTest() function writes a test file to the SD card. The function opens a file for writing, writes some text data to the file, and then closes the file. See Listing 5-59.

Listing 5-59. The WriteFileTest() function

```
void WriteFileTest(String Filename)
{
  File TempFile;
  TempFile = SD.open(Filename.c_str(), FILE_WRITE);
  if (TempFile)
  {
```

```
    Serial.print(F("Writing to testfile ..."));

    TempFile.print(F("TEST CAMERA SDCARD HOOKUP At Time... "));

    TempFile.print(millis()/1000);

    TempFile.println(F(" Seconds"));


    TempFile.print(F("Photo Info Filename: "));

    TempFile.println(CreatePhotoInfoFilename());

    TempFile.print(F("Photo Info:"));

    TempFile.println(CreatePhotoInfo());


    // close the file:

    TempFile.close();

    Serial.println(F("Writing File Done..."));

  }

  else

  {

    // if the file didn't open, print an error:

    Serial.print(F("Error opening "));

    Serial.println(Filename);

  }

}
```

Changing the Camera's Command or Parameters

The ParseRawCommand() function processes the user's camera commands and parameter changes.

The ParseRawCommand() function does the following:

1. Calls the ParseCommand(RawCommandLine.c_str(), ' ', Entries) function that breaks up the user's input into an array of strings.

2. Each of the elements of the string array from Step 1 is sent to the ProcessRawCommandElement(Entries[i]) function to be processed. If the processing was successful then a message is printed out saying that the user has successfully changed a camera command or parameter. Otherwise an error message is printed out.

3. The Resolution is set to "None" which resets and reloads the camera's registers when a photo is taken.
4. The camera's registers are reset by calling the ResetCameraRegisters() function.

See Listing 5-60.

Listing 5-60. The ParseRawCommand() function

```
void ParseRawCommand(String RawCommandLine)
{
  String Entries[10];
  boolean success = false;


  // Parse into command and parameters
  int NumberElements = ParseCommand(RawCommandLine.c_str(), ' ', Entries);


  for (int i = 0 ; i < NumberElements; i++)
  {
    boolean success = ProcessRawCommandElement(Entries[i]);
    if (!success)
    {
      Serial.print(F("Invalid Command or Parameter: "));
      Serial.println(Entries[i]);
    }
    else
    {
      Serial.print(F("Command or parameter "));
      Serial.print(Entries[i]);
      Serial.println(F(" sucessfully set ..."));
    }
  }
}
```

// Assume parameter change since user is setting parameters on command line manually

// Tells the camera to re-initialize and set up camera according to new parameters

Resolution = None; // Reset and reload registers

ResetCameraRegisters();

}

The ParseCommand() function splits an array of characters based on a user input character and stores the result in an array of strings.

The ParseCommand() function does the following:

1. For each character in the commandline compare that character to the splitcharacter if they are equal then this marks the end of a user parameter so save the temp string in the array of strings Result. If the character is not the splitcharacter then add this character to the temp string variable.

2. Add the temp string variable to the array of strings Result. This adds in the last user parameter that does not have a splitcharacter after it to the array of strings.

3. Return the number of user parameters in the Result array.

See Listing 5-61.

Listing 5-61 The ParseCommand() function

```
int ParseCommand(const char* commandline, char splitcharacter, String* Result)
{
  int ResultIndex = 0;

  int length = strlen(commandline);

  String temp = "";


  for (int i = 0; i < length ; i++)
  {
    char tempchar = commandline[i];

    if (tempchar == splitcharacter)
    {
      Result[ResultIndex] += temp;

      ResultIndex++;

      temp = "";
```

```
    }
    else
    {
        temp += tempchar;
    }
}

// Put in end part of string
Result[ResultIndex] = temp;


return (ResultIndex + 1);
}
```

The ProcessRawCommandElement() function tests user input to see if it is one of the camera's commands or camera parameters. If so then it sets the appropriate command or parameter to the user's selection.

The ProcessRawCommandElement() function does the following:

1. Changes the input Element to all lower case in order to help test for valid commands and parameters. For example, the user can enter a command or parameter that is case insensitive such that VGA is the same as vga and is the same as vGa.

2. The input Element is tested to see if it is a valid camera command. Valid commands are vga, vgap, qvga, and qqvga. If there is a match then the command is set to the appropriate command based on the Element.

3. If it is not a command then the input Element is tested to see if it is one of the valid camera parameters. If it is then it is set to the corresponding parameter value based on the Element's value.

4. If the input Element was found it be either a command or parameter then a true is returned otherwise a false value is returned.

See Listing 5-62.

Listing 5-62. The ProcessRawCommandElement() function

```
boolean ProcessRawCommandElement(String Element)
{
    boolean result = false;
```

```
Element.toLowerCase();

if ((Element == "vga") ||
   (Element == "vgap") ||
   (Element == "qvga")||
   (Element == "qqvga"))
{
  Element.toUpperCase();
  Command = Element;
  result = true;
}
else
if (Element == "thirtyfps")
{
  FPSParam = "ThirtyFPS";
  result = true;
}
else
if (Element == "nightmode")
{
  FPSParam = "NightMode";
  result = true;
}
else
if (Element == "sawb")
{
  AWBParam = "SAWB";
  result = true;
}
```

```
else
if (Element == "aawb")
{
  AWBParam = "AAWB";
  result = true;
}
else
if (Element == "aveaec")
{
  AECParam = "AveAEC";
  result = true;
}
else
if (Element == "histaec")
{
  AECParam = "HistAEC";
  result = true;
}
else
if (Element == "yuvmatrixon")
{
  YUVMatrixParam = "YUVMatrixOn";
  result = true;
}
else
if (Element == "yuvmatrixoff")
{
  YUVMatrixParam = "YUVMatrixOff";
  result = true;
```

```
    }
    else
    if (Element == "denoiseyes")
    {
      DenoiseParam = "DenoiseYes";
      result = true;
    }
    else
    if (Element == "denoiseno")
    {
      DenoiseParam = "DenoiseNo";
      result = true;
    }
    else
    if (Element == "edgeyes")
    {
      EdgeParam = "EdgeYes";
      result = true;
    }
    else
    if (Element == "edgeno")
    {
      EdgeParam = "EdgeNo";
      result = true;
    }
    else
    if (Element == "ablcon")
    {
      ABLCParam = "AblcON";
```

```
    result = true;

  }

  else

  if (Element == "ablcoff")

  {

    ABLCParam = "AblcOFF";

    result = true;

  }

  return result;

}
```

Overview of FFMPEG

In order to convert the image files that are saved to the SD card to a viewable picture you need to use an image converter program like ffmpeg. This section gives some background of ffmpeg including the official web site, the version of ffmpeg tested with this book, and the exact commands that will convert the QQVGA, QVGA, and VGA images generated from the capture program into PNG (Portable Network Graphics) image files. The PNG format is widely supported compared to the YUV or Bayer Raw image formats.

Official Website

The official web site for the ffmpeg program is

https://www.ffmpeg.org/

You should go there and download the latest version for your computer system. It is free and available at no charge.

FFMPEG Version Tested with this Book

According to the readme.txt file in the version of ffmpeg I used for this book, the ffmpeg version is the Win32 static build by Kyle Schwarz.

The FFmpeg version is "2015-03-12 git-3bedc99"

This is the version I used to test the output of the ov7670 camera with both the YUV and the Bayer RAW picture files. If you are having problems converting your YUV or RAW image files to a viewable format then try to use this version of ffmpeg. They should have an archive with older versions of the program available.

FFMPEG CONVERSION

This section describes the ffmpeg command line options that are required to convert the images into an easily viewable format. The images from the camera that are recorded in YUV and Raw Bayer format are converted to PNG files. PNG stands for portable network graphics file which is a common image file type that can be viewed in windows explorer or though a paint program like Paint Shop Pro.

For each of the following commands replace INPUTFILE.YUV or INPUTFILE.RAW with the actual filename of the YUV file or Bayer RAW file you want to covert. Change the OUTPUTFILE.PNG filename to the filename you want to save the new image as.

QQVGA

ffmpeg -f rawvideo -s 160x120 -pix_fmt yuyv422 -i INPUTFILE.YUV -f image2 -vcodec png OUTPUTFILE.PNG

QVGA

ffmpeg -f rawvideo -s 320x240 -pix_fmt yuyv422 -i INPUTFILE.YUV -f image2 -vcodec png OUTPUTFILE.PNG

VGA

ffmpeg -f rawvideo -s 640x480 -pix_fmt bayer_bggr8 -i INPUTFILE.RAW -f image2 -vcodec png OUTPUTFILE.PNG

Summary

In this chapter I have covered the hardware and software aspects of the SD card reader including input/output pin connections to the Arduino as well as discussing the SD card itself, and initializing, reading, writing, and deleting files to and from the SD card. Next, I covered the Arduino's I2C interface including the actual interface, reading and writing to an I2C device such as the ov7670 camera. My image capture program that works with the SD card reader and ov7670 camera is also discussed. Finally, the ffmpeg program that is used to convert the YUV and Bayer RAW photo images into PNG image files is covered.

Chapter 6

Taking Photos with the Omnivision ov7670 Camera – Part 2

This chapter is a quick start guide to getting your ov7670 camera operational. I discuss everything you need to know including how to connect the SD card reader and camera to your Arduino Mega 2560. I discuss how to operate the image capture software. I tell you how to transfer your images to your computer and I show you how to convert these images into easily viewable PNG images.

Hands on Example: Taking a picture with the camera, saving the picture to the SD card storage, and viewing the image on your computer.

In this hands on example I will be showing how to setup the camera to take pictures, transfer them to your computer and to convert them to an image format that is easily viewable. I start with showing you how to set up and connect the SD card reader and camera to the Arduino. I then discuss the image capture program I wrote to capture and save images to a SD card. Then, I walk you step by step on how to use the program. I cover how to take a picture in each of the camera command modes that are QQVGA, QVGA, VGAP, and VGA. I also demonstrate how camera parameters can be changed and show how they effect the output of a camera image. I then discuss how to transfer the image data on the SD card to your computer. Finally, I show you how to convert images from the camera to easily viewable images on your computer.

Creating a 3.3 volt Breadboard Node

The first thing you need to do is create a 3.3 volt node on a breadboard that will provide power to the camera and to the SD card reader. To do this you need to connect two male to female jumper wires and 1 wire with male connections on both ends into holes in the breadboard that are connected together. The two male to female wires will be used to connect the 3.3 volt input pins on the SD card reader and camera to the 3.3 volt node on the breadboard. The wire with the male connections on both ends will be used to connect the 3.3 volt node on the breadboard to the 3.3 volt output from the Arduino.

Some breadboards have a power line and a ground line where all the pins in the column are connected. They also have rows of 5 pins that are connected together. See Figure 6-1.

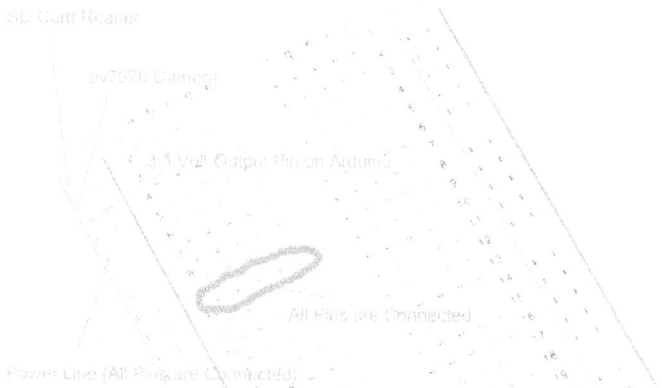

Figure 6-1. Breadboard with power rails and ground line

There are also smaller breadboards available that just have two columns of rows of nodes. Each node consists of 5 pins. I used one of these for the power node for the hands on example in this book. See Figure 6-2. The figure shows the 3.3 volt power node. You can buy both kinds of breadboards on Amazon.com.

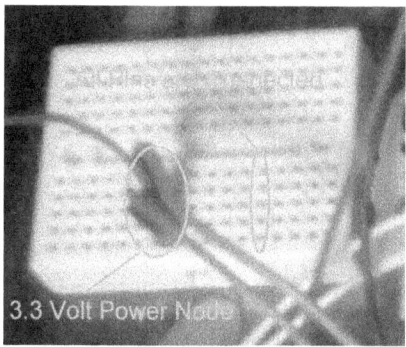

Figure 6-2. Breadboard with only two columns of connected pins

Connecting the SD Card Reader to the Arduino

Next, you need to connect the SD card reader to the Arduino. Make sure the power is off to the Arduino before you perform the following steps. You will need to use the female to male jumper wires for connecting the SD card to the Arduino.

You will need to:

- Connect the 3.3 volt pin on the SD card reader to the breadboard's 3.3 volt node.

- Connect the SDCS pin to pin 48 on the Arduino.

- Connect the MOSI pin to pin 51 on the Arduino.

- Connect the SCK pin to pin 52 on the Arduino.

- Connect the MISO pin to pin 50 on the Arduino.

- Connect the GND pin to a GND pin on the Arduino.

See Figure 6-3.

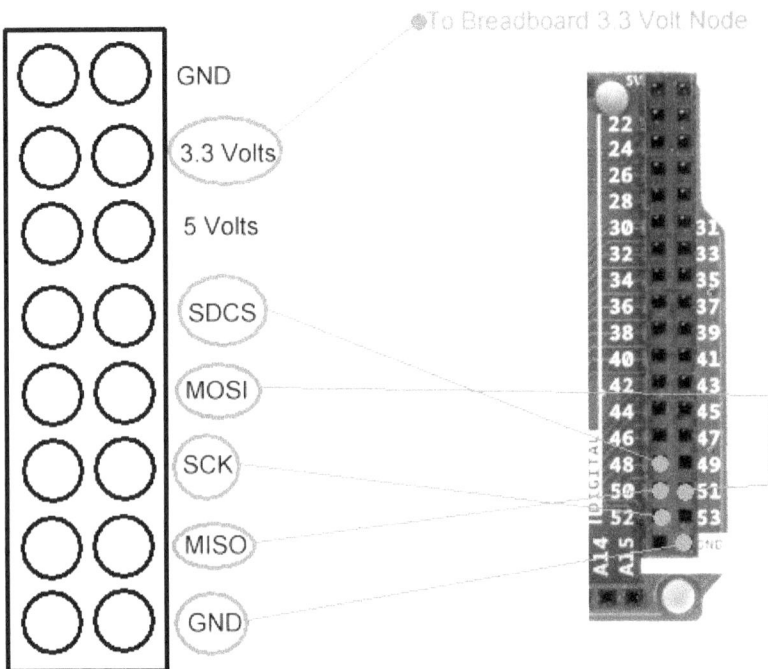

Figure 6-3. Connecting the SD card to the Arduino Mega 2560

Defining the Pins in Code

The user definable pin assignments for a SD card is just the SDCS or chip select pin which is defined in our image capture program as pin 48. This means that the wire from the camera's SDCS pin is connected to the Arduino's pin 48.

const int chipSelect = 48;

Connecting the Camera to the Arduino

Next, we need to connect the ov7670 camera to the Arduino. You will need to use the female to male jumper wires for connecting the camera to the Arduino.

To connect the camera to the Arduino you need to:

- Connect the 3.3 volt pin on the camera to the 3.3 volt node on the breadboard.

- Connect the Ground pin on the camera to one of the Ground pins on the Arduino

- Connect the SIOC pin on the camera to the SCL or pin 21 on the Arduino.
- Connect the SIOD pin on the camera to the SDA or pin 20 on the Arduino.
- Connect the WRST pin on the camera to pin 22 on the Arduino.
- Connect the RRST pin on the camera to pin 23 on the Arduino.
- Connect the WR pin on the camera to pin 24 on the Arduino.
- Connect the VSYNC pin on the camera to pin 25 on the Arduino.
- Connect the RCLK pin on the camera to pin 26 on the Arduino.
- Connect the OE pin on the camera to one of the GND (Ground) pins on the Arduino.
- Connect the D7 pin on the camera to pin 28 on the Arduino.
- Connect the D6 pin on the camera to pin 29 on the Arduino.
- Connect the D5 pin on the camera to pin 30 on the Arduino.
- Connect the D4 pin on the camera to pin 31 on the Arduino.
- Connect the D3 pin on the camera to pin 32 on the Arduino.
- Connect the D2 pin on the camera to pin 33 on the Arduino.
- Connect the D1 pin on the camera to pin 34 on the Arduino.
- Connect the D0 pin on the camera to pin 35 on the Arduino.

See Figures 6-4 and 6-5 for a visual description of the connections required.

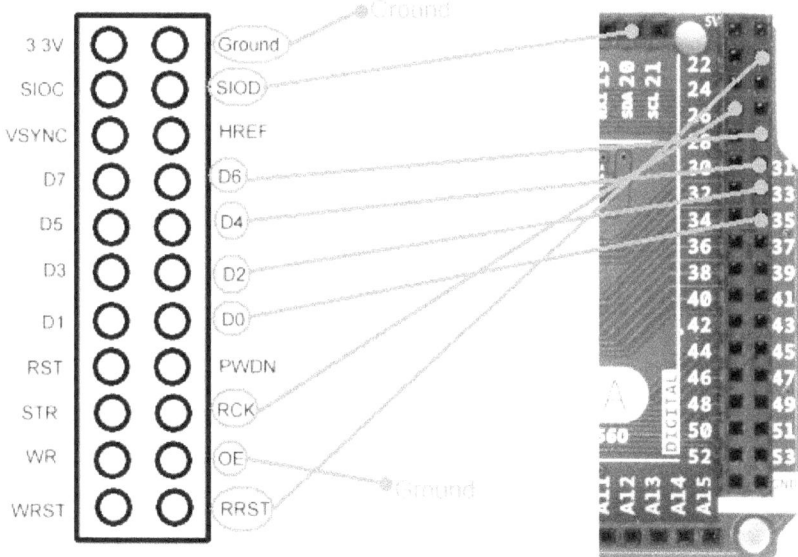

Figure 6-4. Connecting the camera to the Arduino

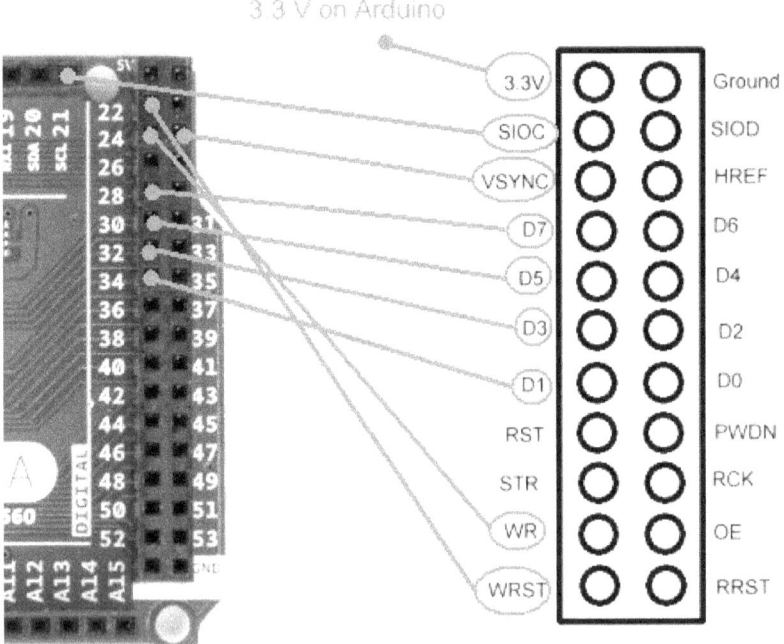

Figure 6-5. Connecting the camera to the Arduino

Defining the Pins in Code

The user definable pins on the Arduino side for the camera are in Listing 6-1. For example, the WRST pin is defined on the Arduino as pin 22 and is connected to the WRST pin on the camera. The RRST pin is defined on the Arduino as pin 23 and is connected to the RRST pin on the camera.

Listing 6-1. User definable pins in our image capture code

```
// Camera input/output pin connection to Arduino

#define WRST  22    // Output Write Pointer Reset

#define RRST  23    // Output Read Pointer Reset

#define WEN   24    // Output Write Enable

#define VSYNC 25    // Input Vertical Sync marking frame capture

#define RCLK  26    // Output FIFO buffer output clock



// FIFO Ram input pins

#define DO7  28

#define DO6  29
```

```
#define DO5  30

#define DO4  31

#define DO3  32

#define DO2  33

#define DO1  34

#define DO0  35
```

The Omnivision ov7670 FIFO Camera Image Capture Software

This section discusses the image capture software. The official web site is listed followed by a listing of the various software commands.

Official Web Site

Currently the official web site of this book is located at:

http://www.psycho-sphere.com

Go to the web site for updates to the image capture program as well as corrections and updates to this book.

A direct link to a zip file of the image capture program is:

http://www.psycho-sphere.com/BeginningArduinoov7670CameraDevelopment.zip

User Camera Control Commands

- d - Display Current Camera Command
- t - Take Photograph using current Command and Parameters

Camera Command and Parameter Set Commands

- Resolution Change Commands: VGA, VGAP, QVGA, QQVGA
- FPS (Frames Per Second) Parameters: ThirtyFPS, NightMode
- AWB (Automatic White Balance) Parameters: SAWB, AAWB
- AEC (Automatic Exposure Control) Parameters: AveAEC, HistAEC
- YUV Color Matrix Parameters: YUVMatrixOn, YUVMatrixOff
- Denoise Parameters: DenoiseYes, DenoiseNo

- Edge Enhancement Parameters: EdgeYes, EdgeNo

- ABLC (Automatic Black Level Calibration) Parameters: AblcON, AblcOFF

SD Card Test Commands

- testread - Tests reading files from the SD card by reading and printing the contents of "test.txt" that was generated by the "testwrite" command. The text output of this command should be something similar to:

TEST.TXT:

TEST CAMERA SDCARD HOOKUP At Time... 25 Seconds

Photo Info Filename: QQVGA0.txt

Photo Info:QQVGA ThirtyFPS SAWB HistAEC YUVMatrixOn DenoiseNo EdgeNo AblcON

- testwrite - Tests writing files to the SD card under the filename "test.txt". The text output from this command is "Writing to testfile ...Writing File Done..."

Help Commands

- help – Displays help menu

- help camera - Displays Camera's Commands and Parameters

Taking Photos with the Camera

The purpose of this section is to demonstrate how to take photos with the ov7670 camera using my image capture software.

The recommended setup of the camera is shown in Figure 6-6 with the camera hanging over the edge of the table so that there is no strain placed on the camera's wires.

> IMPORTANT: Take note of how the camera hangs over the side of the table so that there is no strain on the wires going into the Arduino pins from the camera. I have noticed that when I try to take pictures when I move the camera to a level position and thus put strain on the wires, many times I get an incorrect garbled photo.

Figure 6-6. Recommended camera setup

Image Capture Program Initialization

1. First you will need to download and uncompress the image capture program using a program like 7-zip.

    Note: 7 zip is a free program that is available at: http://www.7-zip.org

2. Start the Arduino IDE and load in the image capture program you just downloaded and uncompressed.

3. Connect your Arduino Mega 2560 to your computer. Also make sure the SD card is pushed firmly into the SD card reader's slot until there is a click. It is also a good idea to check that the SD card reader is working properly by using the "testread" and "testwrite" commands. You should first use "testwrite" to have the Arduino write out a file called "test.txt" to the SD card. Then, use the "testread" command to read back the file and print its contents out to the screen.

4. Click the "Upload" button on the Arduino IDE to upload the image capture program to your Arduino.

5. Wait until the program has finished uploading.

6. Click the "Serial Monitor" button to bring up the serial monitor.

7. The program should begin to initialize and produce the following output on the Serial Monitor window.

Arduino SERIAL_MONITOR_CONTROLLED CAMERA ... Using ov7670 Camera

---------------------------- Camera Registers ----------------------------

RESETTING ALL REGISTERS BY SETTING COM7 REGISTER to 0x80: I2C Operation OK ...

CLKRC = 80

COM7 = 0

COM3 = 0

COM14 = 0

SCALING_XSC = 3A

SCALING_YSC = 35

SCALING_DCWCTR = 11

SCALING_PCLK_DIV = 0

SCALING_PCLK_DELAY = 2

TSLB (YUV Order- Higher Bit, Bit[3]) = D

COM13 (YUV Order - Lower Bit, Bit[1]) = 88

COM17 (DSP Color Bar Selection) = 0

COM4 (works with COM 17) = 0

COM15 (COLOR FORMAT SELECTION) = C0

COM11 (Night Mode) = 0

COM8 (Color Control, AWB) = 8F

HAECC7 (AEC Algorithm Selection) = 14

GFIX = 0

HSTART = 11

HSTOP = 61

HREF = 80

VSTRT = 3

VSTOP = 7B

VREF = 0

In SetupCamera() ...

Initializing OV7670 Camera ...

FINISHED INITIALIZING CAMERA ...

Initializing SD card...Wiring is correct and a card is present ...

Omnivision ov7670 Camera Image Capture Software Version 1.0

Copyright 2015 by Robert Chin. All Rights Reserved.

Type h or help for main help menu ...

Ready to Accept new Command =>

Taking QQVGA Photos

8. Now lets take a photo by sending a "t" to the Arduino via the Serial Monitor. Since the default is set to take pictures in QQVGA mode the picture that is taken will be of QQVGA resolution. The following should appear in the output window. Figure 6-7 and Figure 6-8 shows you some photos that I took in QQVGA format.

    Important Note: The default focus of the camera may need adjustment. After taking this first picture transfer the image to your computer and then convert it into an easily viewable image file using ffmpeg (please see the section on ffmpeg in this chapter). Look at the image and see if it is in focus. If it is not then you will need to adjust the camera's focus by turning the lens. Take another picture. If the new picture is clearer than the old one then you are turning the lens in the right direction. If it is less focused then turn the camera's lens in the opposite direction. Continue taking photos, checking the sharpness of the picture, and then adjusting the camera lens until you get a clear picture.

Ready to Accept new Command =>

Raw Command from Serial Monitor: t

Going to take photo with current command:

Current Command:

Command: QQVGA

FPSParam: ThirtyFPS

AWBParam: SAWB

AECParam: HistAEC

YUVMatrixParam: YUVMatrixOn

DenoiseParam: DenoiseNo

EdgeParam: EdgeNo

ABLCParam: AblcON

Taking a QQVGA Photo...

-------------------------- Setting Camera for QQVGA YUV --------------------------

Photo Width = 160

Photo Height = 120

Bytes Per Pixel = 2

.... Setting Basic QQVGA Parameters ....

CLKRC: I2C Operation OK ...

COM7: I2C Operation OK ...

COM3: I2C Operation OK ...

COM14: I2C Operation OK ...

SCALING_XSC: I2C Operation OK ...

SCALING_YSC: I2C Operation OK ...

SCALING_DCWCTR: I2C Operation OK ...

SCALING_PCLK_DIV: I2C Operation OK ...

SCALING_PCLK_DELAY: I2C Operation OK ...

TSLB: I2C Operation OK ...

COM13: I2C Operation OK ...

COM17: I2C Operation OK ...

.... Setting Camera to 30 FPS ....

CLKRC: I2C Operation OK ...

DBLV: I2C Operation OK ...

EXHCH: I2C Operation OK ...

EXHCL: I2C Operation OK ...

DM_LNL: I2C Operation OK ...

DM_LNH: I2C Operation OK ...

COM11: I2C Operation OK ...

-------------- Setting Camera Histogram Based AEC/AGC Registers --------------

AEW: I2C Operation OK ...

AEB: I2C Operation OK ...

HAECC1: I2C Operation OK ...

HAECC2: I2C Operation OK ...

HAECC3: I2C Operation OK ...

HAECC4: I2C Operation OK ...

HAECC5: I2C Operation OK ...

HAECC6: I2C Operation OK ...

HAECC7: I2C Operation OK ...

.... Setting Camera to Simple AWB ....

COM8(0x13): I2C Operation OK ...

AWBCTR0 Control Register 0(0x6F): I2C Operation OK ...

.... Setting Camera Gain ....

COM9: I2C Operation OK ...

BLUE GAIN: I2C Operation OK ...

RED GAIN: I2C Operation OK ...

GREEN GAIN: I2C Operation OK ...

COM16(ENABLE GAIN): I2C Operation OK ...

.... Setting Camera Undocumented Registers ....

Setting B0 UNDOCUMENTED register to 0x84:= I2C Operation OK ...

.... Setting Camera Color Matrix for YUV ....

MTX1: I2C Operation OK ...

MTX2: I2C Operation OK ...

MTX3: I2C Operation OK ...

MTX4: I2C Operation OK ...

MTX5: I2C Operation OK ...

MTX6: I2C Operation OK ...

CONTRAS: I2C Operation OK ...

MTXS: I2C Operation OK ...

.... Setting Camera Saturation Level ....

SATCTR: I2C Operation OK ...

.... Setting Camera Array Control ....

CHLF: I2C Operation OK ...

ARBLM: I2C Operation OK ...

.... Setting Camera ADC Control ....

ADCCTR1: I2C Operation OK ...

ADCCTR2: I2C Operation OK ...

ADC: I2C Operation OK ...

ACOM: I2C Operation OK ...

OFON: I2C Operation OK ...

.... Setting Camera ABLC ....

ABLC1: I2C Operation OK ...

THL_ST: I2C Operation OK ...

.... Setting Camera Window Output Parameters ....

HSTART: I2C Operation OK ...

HSTOP: I2C Operation OK ...

HREF: I2C Operation OK ...

VSTRT: I2C Operation OK ...

VSTOP: I2C Operation OK ...

VREF: I2C Operation OK ...

---------------------------- Camera Registers ---------------------------

CLKRC = 80

COM7 = 0

COM3 = 4

COM14 = 1A

SCALING_XSC = 3A

SCALING_YSC = 35

SCALING_DCWCTR = 22

SCALING_PCLK_DIV = F2

SCALING_PCLK_DELAY = 2

TSLB (YUV Order- Higher Bit, Bit[3]) = 0

COM13 (YUV Order - Lower Bit, Bit[1]) = C8

COM17 (DSP Color Bar Selection) = 0

COM4 (works with COM 17) = 0

COM15 (COLOR FORMAT SELECTION) = C0

COM11 (Night Mode) = A

COM8 (Color Control, AWB) = E7

HAECC7 (AEC Algorithm Selection) = 94

GFIX = 0

HSTART = 13

HSTOP = 1

HREF = A4

VSTRT = 2

VSTOP = 7A

VREF = A

Starting Capture of Photo ...

Time for Frame Capture (milliseconds) = 130

VSync beginning duration (microseconds) = 392

VSync end duration (microseconds) = 392

Starting Transmission of Photo To SDCard ...

Total Bytes Saved to SDCard = 38400

Writing Photo's Info file (.txt file) to SD Card ...

Elapsed Time for Taking and Sending Photo(secs) = 4

Photo Taken and Saved to Arduino SD CARD ...

Image Output Filename :QQVGA0.yuv

Ready to Accept new Command =>

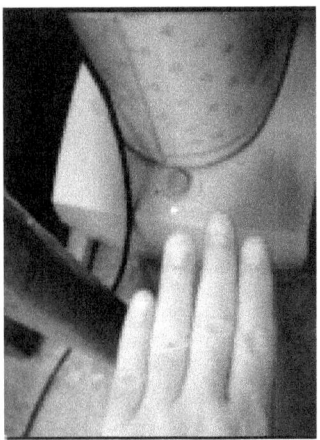

Figure 6-7. QQVGA Photo

Figure 6-8. QQVGA Photo

Taking QVGA Photos

9. Next, let's take some QVGA photos. Send the value "qvga" to the Arduino using the Serial Monitor. The command should be ackowledged, camera registers should be reset, and the new camera command with parameters should be displayed. You should see output similar to the following:

Ready to Accept new Command =>

Raw Command from Serial Monitor: qvga

Changing command or parameters according to your input:

Command or parameter qvga sucessfully set ...

RESETTING ALL REGISTERS BY SETTING COM7 REGISTER to 0x80: I2C Operation OK ...

Current Command:

Command: QVGA

FPSParam: ThirtyFPS

AWBParam: SAWB

AECParam: HistAEC

YUVMatrixParam: YUVMatrixOn

DenoiseParam: DenoiseNo

EdgeParam: EdgeNo

ABLCParam: AblcON

10. To take a photo send the Arduino a "t" value. The command should be acknowledged, the current camera command with parameters should be displayed, a detailed breakdown of the exact camera register write operations being performed is displayed, a printout of the resulting changes in key registers and information regarding the saving of the photo to the SD Card is displayed. Some photos that I have taken are shown in Figure 6-9 and Figure 6-10. You should see output similar to the following:

Ready to Accept new Command =>

Raw Command from Serial Monitor: t

Going to take photo with current command:

Current Command:

Command: QVGA

FPSParam: ThirtyFPS

AWBParam: SAWB

AECParam: HistAEC

YUVMatrixParam: YUVMatrixOn

DenoiseParam: DenoiseNo

EdgeParam: EdgeNo

ABLCParam: AblcON

Taking a QVGA Photo...

--------------------------- Setting Camera for QVGA (YUV) ---------------------------

Photo Width = 320

Photo Height = 240

Bytes Per Pixel = 2

CLKRC: I2C Operation OK ...

COM7: I2C Operation OK ...

COM3: I2C Operation OK ...

COM14: I2C Operation OK ...

SCALING_XSC: I2C Operation OK ...

SCALING_YSC: I2C Operation OK ...

SCALING_DCWCTR: I2C Operation OK ...

SCALING_PCLK_DIV: I2C Operation OK ...

SCALING_PCLK_DELAY: I2C Operation OK ...

TSLB: I2C Operation OK ...

COM13: I2C Operation OK ...

COM17: I2C Operation OK ...

.... Setting Camera to 30 FPS ....

CLKRC: I2C Operation OK ...

DBLV: I2C Operation OK ...

EXHCH: I2C Operation OK ...

EXHCL: I2C Operation OK ...

DM_LNL: I2C Operation OK ...

DM_LNH: I2C Operation OK ...

COM11: I2C Operation OK ...

-------------- Setting Camera Histogram Based AEC/AGC Registers ---------------

AEW: I2C Operation OK ...

AEB: I2C Operation OK ...

HAECC1: I2C Operation OK ...

HAECC2: I2C Operation OK ...

HAECC3: I2C Operation OK ...

HAECC4: I2C Operation OK ...

HAECC5: I2C Operation OK ...

HAECC6: I2C Operation OK ...

HAECC7: I2C Operation OK ...

.... Setting Camera to Simple AWB ....

COM8(0x13): I2C Operation OK ...

AWBCTR0 Control Register 0(0x6F): I2C Operation OK ...

.... Setting Camera Gain ....

COM9: I2C Operation OK ...

BLUE GAIN: I2C Operation OK ...

RED GAIN: I2C Operation OK ...

GREEN GAIN: I2C Operation OK ...

COM16(ENABLE GAIN): I2C Operation OK ...

.... Setting Camera Undocumented Registers ....

Setting B0 UNDOCUMENTED register to 0x84:= I2C Operation OK ...

.... Setting Camera Color Matrix for YUV ....

MTX1: I2C Operation OK ...

MTX2: I2C Operation OK ...

MTX3: I2C Operation OK ...

MTX4: I2C Operation OK ...

MTX5: I2C Operation OK ...

MTX6: I2C Operation OK ...

CONTRAS: I2C Operation OK ...

MTXS: I2C Operation OK ...

.... Setting Camera Saturation Level ....

SATCTR: I2C Operation OK ...

.... Setting Camera Array Control ....

CHLF: I2C Operation OK ...

ARBLM: I2C Operation OK ...

.... Setting Camera ADC Control ....

ADCCTR1: I2C Operation OK ...

ADCCTR2: I2C Operation OK ...

ADC: I2C Operation OK ...

ACOM: I2C Operation OK ...

OFON: I2C Operation OK ...

.... Setting Camera ABLC ....

ABLC1: I2C Operation OK ...

THL_ST: I2C Operation OK ...

.... Setting Camera Window Output Parameters ....

HSTART: I2C Operation OK ...

HSTOP: I2C Operation OK ...

HREF: I2C Operation OK ...

VSTRT: I2C Operation OK ...

VSTOP: I2C Operation OK ...

VREF: I2C Operation OK ...

--------------------------- Camera Registers ---------------------------

CLKRC = 80

COM7 = 0

COM3 = 4

COM14 = 19

SCALING_XSC = 3A

SCALING_YSC = 35

SCALING_DCWCTR = 11

SCALING_PCLK_DIV = F1

SCALING_PCLK_DELAY = 2

TSLB (YUV Order- Higher Bit, Bit[3]) = 4

COM13 (YUV Order - Lower Bit, Bit[1]) = C2

COM17 (DSP Color Bar Selection) = 0

COM4 (works with COM 17) = 0

COM15 (COLOR FORMAT SELECTION) = C0

COM11 (Night Mode) = A

COM8 (Color Control, AWB) = E7

HAECC7 (AEC Algorithm Selection) = 94

GFIX = 0

HSTART = 13

HSTOP = 1

HREF = 24

VSTRT = 2

VSTOP = 7A

VREF = A

Starting Capture of Photo ...

Time for Frame Capture (milliseconds) = 74

VSync beginning duration (microseconds) = 392

VSync end duration (microseconds) = 392

Starting Transmission of Photo To SDCard ...

Total Bytes Saved to SDCard = 153600

Writing Photo's Info file (.txt file) to SD Card ...

Elapsed Time for Taking and Sending Photo(secs) = 15

Photo Taken and Saved to Arduino SD CARD ...

Image Output Filename :QVGA3.yuv

Ready to Accept new Command =>

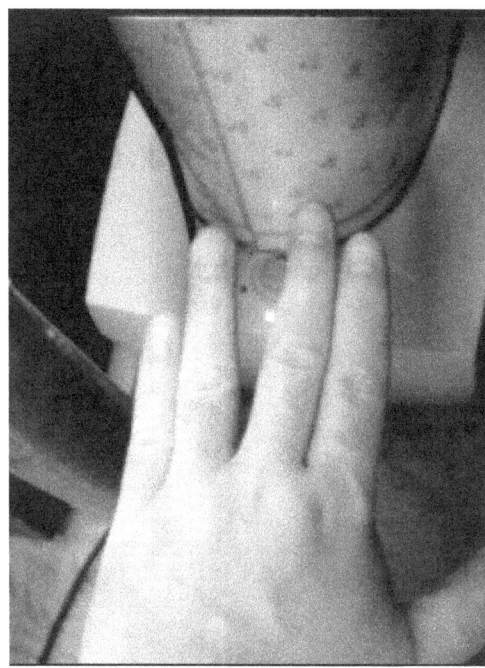

Figure 6-9. QVGA photo

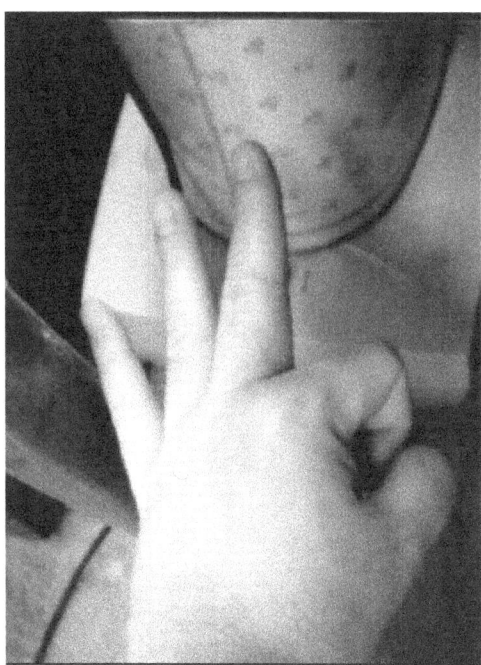

Figure 6-10. QVGA photo

11. Next lets try to change a camera parameter by turning the YUV Color Matrix off. The YUV color matrix has the effect of making the colors more vibrant by increasing the red, green, and blue colors in an image. By turning this matrix off we make the image less vibrant and colorful. Enter "yuvmatrixoff" and send this value to the Arduino via the Serial Monitor. You should see text output like the following which confirms the parameter change, resets the camera's registers, and displays the new changed camera settings:

Ready to Accept new Command =>

Raw Command from Serial Monitor: yuvmatrixoff

Changing command or parameters according to your input:

Command or parameter yuvmatrixoff sucessfully set ...

RESETTING ALL REGISTERS BY SETTING COM7 REGISTER to 0x80: I2C Operation OK ...

Current Command:

Command: QVGA

FPSParam: ThirtyFPS

AWBParam: SAWB

AECParam: HistAEC

YUVMatrixParam: YUVMatrixOff

DenoiseParam: DenoiseNo

EdgeParam: EdgeNo

ABLCParam: AblcON

12. Next, take a photo by sending the Arduino a "t" command. You should see text output similar to the last time you took a photo however the "........... Setting Camera Color Matrix for YUV ...." section should be missing. This indicates that the YUV Color Matrix has been deactivated. See Figure 6-11 for a photo I took with these new parameters.

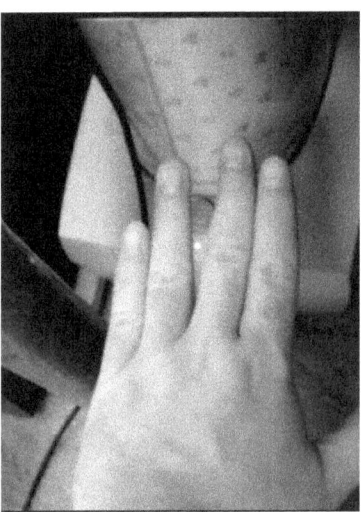

Figure 6-11. QVGA photo with YUV color matrix is turned off

Taking VGAP Photos

13. Next, lets try to take a photo using the processed VGA mode. Send the command "vgap" to the Arduino to change the camera command settings. The command should be confirmed, the camera's registers should be reset, and the current command should be displayed. You should see something like the following:

Ready to Accept new Command =>

Raw Command from Serial Monitor: vgap

Changing command or parameters according to your input:

Command or parameter vgap sucessfully set ...

RESETTING ALL REGISTERS BY SETTING COM7 REGISTER to 0x80: I2C Operation OK ...

Current Command:

Command: VGAP

FPSParam: ThirtyFPS

AWBParam: SAWB

AECParam: HistAEC

YUVMatrixParam: YUVMatrixOff

DenoiseParam: DenoiseNo

EdgeParam: EdgeNo

ABLCParam: AblcON

14. Now, send the Arduino the "t" command to actually take the photo. The current camera settings should be displayed, a detailed list of register setting operations should be displayed, the contents of key registers are displayed, and SD card related information is displayed. See Figure 6-12 for a photo I took using these camera settings. You should see output similar to the following:

Ready to Accept new Command =>

Raw Command from Serial Monitor: t

Going to take photo with current command:

Current Command:

Command: VGAP

FPSParam: ThirtyFPS

AWBParam: SAWB

AECParam: HistAEC

YUVMatrixParam: YUVMatrixOff

DenoiseParam: DenoiseNo

EdgeParam: EdgeNo

ABLCParam: AblcON

Taking a VGAP Photo...

RESETTING ALL REGISTERS BY SETTING COM7 REGISTER to 0x80: I2C Operation OK ...

-------------------------- Setting Camera for VGA (Raw RGB) --------------------------

Photo Width = 640

Photo Height = 480

Bytes Per Pixel = 1

CLKRC: I2C Operation OK ...

COM7: I2C Operation OK ...

COM3: I2C Operation OK ...

COM14: I2C Operation OK ...

SCALING_XSC: I2C Operation OK ...

SCALING_YSC: I2C Operation OK ...

SCALING_DCWCTR: I2C Operation OK ...

SCALING_PCLK_DIV: I2C Operation OK ...

SCALING_PCLK_DELAY: I2C Operation OK ...

COM17: I2C Operation OK ...

.... Setting Camera to 30 FPS ....

CLKRC: I2C Operation OK ...

DBLV: I2C Operation OK ...

EXHCH: I2C Operation OK ...

EXHCL: I2C Operation OK ...

DM_LNL: I2C Operation OK ...

DM_LNH: I2C Operation OK ...

COM11: I2C Operation OK ...

-------------- Setting Camera Histogram Based AEC/AGC Registers ---------------

AEW: I2C Operation OK ...

AEB: I2C Operation OK ...

HAECC1: I2C Operation OK ...

HAECC2: I2C Operation OK ...

HAECC3: I2C Operation OK ...

HAECC4: I2C Operation OK ...

HAECC5: I2C Operation OK ...

HAECC6: I2C Operation OK ...

HAECC7: I2C Operation OK ...

Setting B0 UNDOCUMENTED register to 0x84:= I2C Operation OK ...

.... Setting Camera Saturation Level ....

SATCTR: I2C Operation OK ...

.... Setting Camera Array Control ....

CHLF: I2C Operation OK ...

ARBLM: I2C Operation OK ...

.... Setting Camera ADC Control ....

ADCCTR1: I2C Operation OK ...

ADCCTR2: I2C Operation OK ...

ADC: I2C Operation OK ...

ACOM: I2C Operation OK ...

OFON: I2C Operation OK ...

.... Setting Camera ABLC ....

ABLC1: I2C Operation OK ...

THL_ST: I2C Operation OK ...

.... Setting Camera Window Output Parameters ....

HSTART: I2C Operation OK ...

HSTOP: I2C Operation OK ...

HREF: I2C Operation OK ...

VSTRT: I2C Operation OK ...

VSTOP: I2C Operation OK ...

VREF: I2C Operation OK ...

------------- Setting Camera for VGA (Processed Bayer RGB) ----------------

COM7: I2C Operation OK ...

Initializing TSLB register result = I2C Operation OK ...

Setting B0 UNDOCUMENTED register to 0x84:= I2C Operation OK ...

.... Setting Camera to Simple AWB ....

COM8(0x13): I2C Operation OK ...

AWBCTR0 Control Register 0(0x6F): I2C Operation OK ...

.... Setting Camera Gain ....

COM9: I2C Operation OK ...

BLUE GAIN: I2C Operation OK ...

RED GAIN: I2C Operation OK ...

GREEN GAIN: I2C Operation OK ...

COM16(ENABLE GAIN): I2C Operation OK ...

----------------------------- Camera Registers ---------------------------

CLKRC = 80

COM7 = 5

COM3 = 0

COM14 = 0

SCALING_XSC = 3A

SCALING_YSC = 35

SCALING_DCWCTR = 11

SCALING_PCLK_DIV = F0

SCALING_PCLK_DELAY = 2

TSLB (YUV Order- Higher Bit, Bit[3]) = 4

COM13 (YUV Order - Lower Bit, Bit[1]) = 88

COM17 (DSP Color Bar Selection) = 0

COM4 (works with COM 17) = 0

COM15 (COLOR FORMAT SELECTION) = C0

COM11 (Night Mode) = A

COM8 (Color Control, AWB) = E7

HAECC7 (AEC Algorithm Selection) = 94

GFIX = 0

HSTART = 13

HSTOP = 1

HREF = B6

VSTRT = 2

VSTOP = 7A

VREF = A

Starting Capture of Photo ...

Time for Frame Capture (milliseconds) = 87

VSync beginning duration (microseconds) = 385

VSync end duration (microseconds) = 385

Starting Transmission of Photo To SDCard ...

Total Bytes Saved to SDCard = 307200

Writing Photo's Info file (.txt file) to SD Card ...

Elapsed Time for Taking and Sending Photo(secs) = 30

Photo Taken and Saved to Arduino SD CARD ...

Image Output Filename :VGAP7.raw

Figure 6-12. VGAP mode photo

15. Next lets change the camera's parameters so that edge enhancement is on. What this does is increase the sharpness of the final photo. Send the value "edgeyes" to the Arduino to activate the edge enhancement. The command should be confirmed, the registers reset, and the current camera setting displayed. The output should look like the following:

Ready to Accept new Command =>

Raw Command from Serial Monitor: edgeyes

Changing command or parameters according to your input:

Command or parameter edgeyes sucessfully set ...

RESETTING ALL REGISTERS BY SETTING COM7 REGISTER to 0x80: I2C Operation OK ...

Current Command:

Command: VGAP

FPSParam: ThirtyFPS

AWBParam: SAWB

AECParam: HistAEC

YUVMatrixParam: YUVMatrixOff

DenoiseParam: DenoiseNo

EdgeParam: EdgeYes

ABLCParam: AblcON

16. Next, take the photo by sending the Arduino the "t" value. The text output should be similar to previous examples. See Figure 6-13 for the photo I took using these camera settings. The key thing to note here is that there should be a section that shows that that edge enhancement is now active such as shown in the following:

.... Setting Camera Edge Enhancement ....

EDGE: I2C Operation OK ...

REG75: I2C Operation OK ...

REG76: I2C Operation OK ...

COM16: I2C Operation OK ...

Figure 6-13. VGA photo with edge enhancement on

Taking VGA Photos

17. Next, switch to VGA by sending "vga" to the Arudino. The command should be confirmed, the camera registers should be reset and the current camera settings should be displayed as in the following output:

Ready to Accept new Command =>

Raw Command from Serial Monitor: vga

Changing command or parameters according to your input:

Command or parameter vga sucessfully set ...

RESETTING ALL REGISTERS BY SETTING COM7 REGISTER to 0x80: I2C Operation OK ...

Current Command:

Command: VGA

FPSParam: ThirtyFPS

AWBParam: SAWB

AECParam: HistAEC

YUVMatrixParam: YUVMatrixOff

DenoiseParam: DenoiseNo

EdgeParam: EdgeYes

ABLCParam: AblcON

18. Next, take a photo by sending a "t" character to the Arduino. The current camera settings should be displayed, the camera registers are set to the required values, the values of key camera registers are displayed and photo file information is displayed. See 6-14 for a photo I took using these camera settings. The text output should look similar to the following:

Ready to Accept new Command =>

Raw Command from Serial Monitor: t

Going to take photo with current command:

Current Command:

Command: VGA

FPSParam: ThirtyFPS

AWBParam: SAWB

AECParam: HistAEC

YUVMatrixParam: YUVMatrixOff

DenoiseParam: DenoiseNo

EdgeParam: EdgeYes

ABLCParam: AblcON

Taking a VGA Photo...

RESETTING ALL REGISTERS BY SETTING COM7 REGISTER to 0x80: I2C Operation OK ...

-------------------------- Setting Camera for VGA (Raw RGB) --------------------------

Photo Width = 640

Photo Height = 480

Bytes Per Pixel = 1

CLKRC: I2C Operation OK ...

COM7: I2C Operation OK ...

COM3: I2C Operation OK ...

COM14: I2C Operation OK ...

SCALING_XSC: I2C Operation OK ...

SCALING_YSC: I2C Operation OK ...

SCALING_DCWCTR: I2C Operation OK ...

SCALING_PCLK_DIV: I2C Operation OK ...

SCALING_PCLK_DELAY: I2C Operation OK ...

COM17: I2C Operation OK ...

.... Setting Camera to 30 FPS ....

CLKRC: I2C Operation OK ...

DBLV: I2C Operation OK ...

EXHCH: I2C Operation OK ...

EXHCL: I2C Operation OK ...

DM_LNL: I2C Operation OK ...

DM_LNH: I2C Operation OK ...

COM11: I2C Operation OK ...

-------------- Setting Camera Histogram Based AEC/AGC Registers --------------

AEW: I2C Operation OK ...

AEB: I2C Operation OK ...

HAECC1: I2C Operation OK ...

HAECC2: I2C Operation OK ...

HAECC3: I2C Operation OK ...

HAECC4: I2C Operation OK ...

HAECC5: I2C Operation OK ...

HAECC6: I2C Operation OK ...

HAECC7: I2C Operation OK ...

Setting B0 UNDOCUMENTED register to 0x84:= I2C Operation OK ...

.... Setting Camera Saturation Level ....

SATCTR: I2C Operation OK ...

.... Setting Camera Array Control ....

CHLF: I2C Operation OK ...

ARBLM: I2C Operation OK ...

.... Setting Camera ADC Control ....

ADCCTR1: I2C Operation OK ...

ADCCTR2: I2C Operation OK ...

ADC: I2C Operation OK ...

ACOM: I2C Operation OK ...

OFON: I2C Operation OK ...

.... Setting Camera ABLC ....

ABLC1: I2C Operation OK ...

THL_ST: I2C Operation OK ...

.... Setting Camera Window Output Parameters ....

HSTART: I2C Operation OK ...

HSTOP: I2C Operation OK ...

HREF: I2C Operation OK ...

VSTRT: I2C Operation OK ...

VSTOP: I2C Operation OK ...

VREF: I2C Operation OK ...

---------------------------- Camera Registers ---------------------------

CLKRC = 80

COM7 = 1

COM3 = 0

COM14 = 0

SCALING_XSC = 3A

SCALING_YSC = 35

SCALING_DCWCTR = 11

SCALING_PCLK_DIV = F0

SCALING_PCLK_DELAY = 2

TSLB (YUV Order- Higher Bit, Bit[3]) = D

COM13 (YUV Order - Lower Bit, Bit[1]) = 88

COM17 (DSP Color Bar Selection) = 0

COM4 (works with COM 17) = 0

COM15 (COLOR FORMAT SELECTION) = C0

COM11 (Night Mode) = A

COM8 (Color Control, AWB) = 8F

HAECC7 (AEC Algorithm Selection) = 94

GFIX = 0

HSTART = 13

HSTOP = 1

HREF = B6

VSTRT = 2

VSTOP = 7A

VREF = A

Starting Capture of Photo ...

Time for Frame Capture (milliseconds) = 90

VSync beginning duration (microseconds) = 392

VSync end duration (microseconds) = 392

Starting Transmission of Photo To SDCard ...

Total Bytes Saved to SDCard = 307200

Writing Photo's Info file (.txt file) to SD Card ...

Elapsed Time for Taking and Sending Photo(secs) = 30

Photo Taken and Saved to Arduino SD CARD ...

Image Output Filename :VGA12.raw

Note: Since VGA mode does not include digital signal processing on the image, features like edge enhancement that use the Digital Signal Processor in the camera will not work.

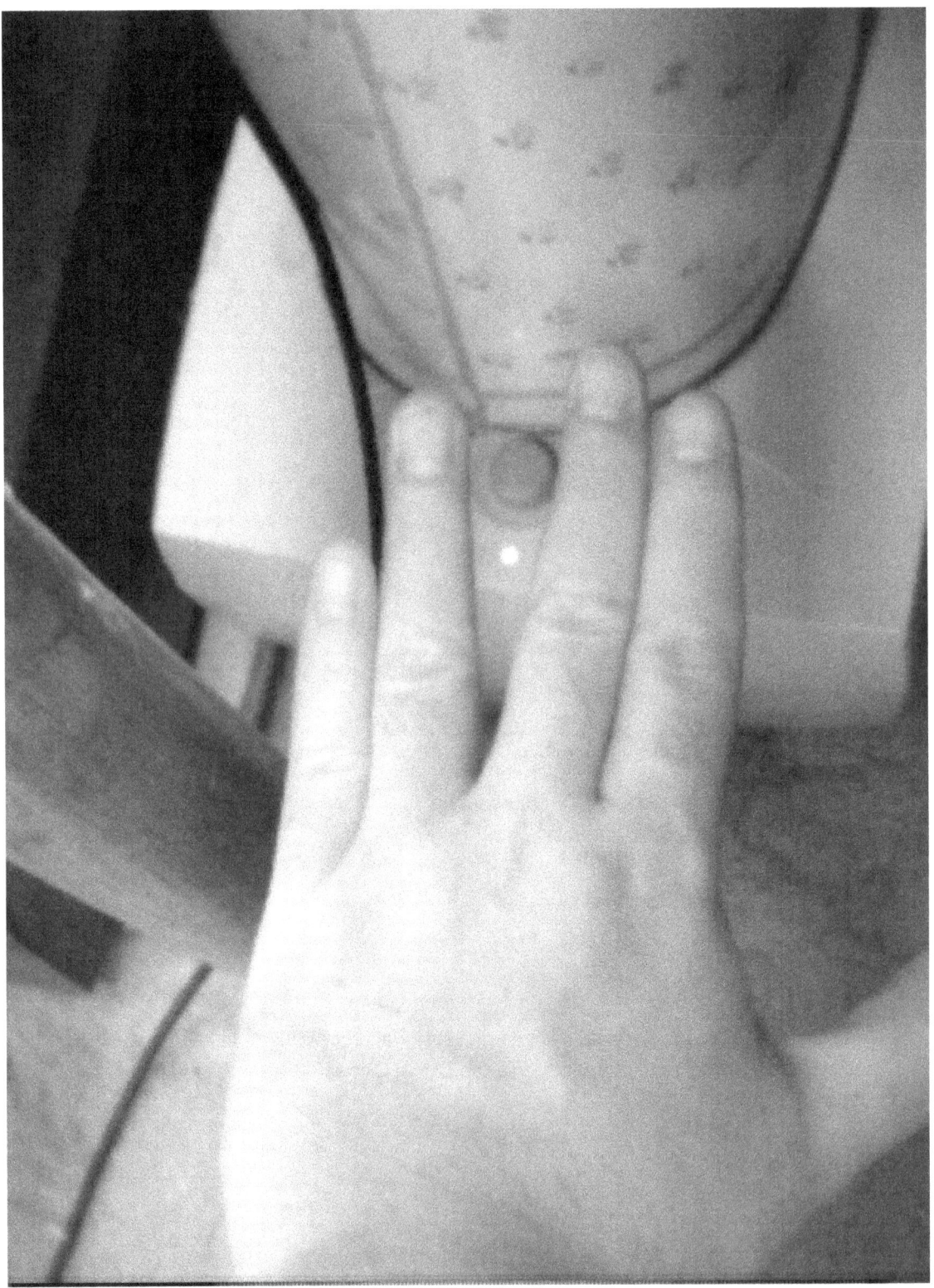

Figure 6-14. VGA photo

Transferring the Photos to your Computer

In order to transfer your photos from the SD card to your computer for final processing you will need another SD card reader designed to work with your computer.

I purchased the "Transcend Information USB 3.0 Card Reader (TS-RDF5K)" from Amazon.com to transfer my files to my Windows based PC. See Figure 6-15. There were some reviews complaining of overheating after use but I did not encounter any problems with excessive heating of the SD card reader.

You will need to stick your SD card into the slot marked as SD. Then stick the SD card reader into an available USB slot on your computer. The computer should recognize the SD card reader as just another disk drive. You can use the Windows Explorer to copy files from the SD card to a folder on your hard disk drive.

Figure 6-15. SD card reader for PC

Converting the Photos to PNG (Portable Network Graphics) Format

Once you get your files to your computer you will need to convert these files to a form that is easily readable and viewable such as the PNG image format. To do this you will need a program called ffmpeg.

The official web site for the ffmpeg program is

https://www.ffmpeg.org/

You should go there and download the latest version for your computer system. It is free and available at no charge.

> Note: The ffmpeg version I used for this book is "2015-03-12 git-3bedc99". If you are having problems converting the camera images into a viewable PNG file then try to download this version of ffmpeg.

FFMPEG CONVERSION

This section describes the ffmpeg command line options that are required to convert the images into an easily viewable format. The images from the camera that are recorded in YUV and Raw Bayer format are converted to PNG files. PNG stands for portable network graphics file which is a common image file type that can be viewed in Windows Explorer or though a paint program like Paint Shop Pro.

> Note: For each of the following commands replace INPUTFILE.YUV or INPUTFILE.RAW with the actual filename of the YUV file or Bayer RAW file you want to covert. Change the OUTPUTFILE.PNG filename to the filename you want to save the new image as.

QQVGA

ffmpeg -f rawvideo -s 160x120 -pix_fmt yuyv422 -i INPUTFILE.YUV   -f image2 -vcodec png OUTPUTFILE.PNG

QVGA

ffmpeg -f rawvideo -s 320x240 -pix_fmt yuyv422 -i INPUTFILE.YUV   -f image2 -vcodec png OUTPUTFILE.PNG

VGA

ffmpeg -f rawvideo -s 640x480 -pix_fmt bayer_bggr8 -i INPUTFILE.RAW   -f image2 -vcodec png OUTPUTFILE.PNG

An easy way to convert the files is to copy the image files into the same directory where your ffmpeg.exe is located. Then you can create a directory such as "outputdir" to hold the processed images. Based on the image filename you can choose one of the commands below to convert the file and to place the converted image file into a directory.

> Note: Replace the "outputdir" name with the name of the directory that you want to save the converted file in.

QQVGA

ffmpeg -f rawvideo -s 160x120 -pix_fmt yuyv422 -i INPUTFILE.YUV   -f image2 -vcodec png outputdir\OUTPUTFILE.PNG

QVGA

ffmpeg -f rawvideo -s 320x240 -pix_fmt yuyv422 -i INPUTFILE.YUV   -f image2 -vcodec png outputdir\OUTPUTFILE.PNG

VGA

ffmpeg -f rawvideo -s 640x480 -pix_fmt bayer_bggr8 -i INPUTFILE.RAW   -f image2 -vcodec png outputdir\OUTPUTFILE.PNG

Summary

This chapter provided a quick start guide to taking photos with the ov7670 camera and saving them using a SD card. SD card reader connections and camera connections to the Arduino are shown. The image capture software is discussed and step by step instructions on how to use it are given. Then I show how to transfer these photos to your computer and then convert them into viewable images.

Appendix A: Camera Register Defines

This appendix lists the #defines related to the camera registers.

// Register addresses and values

#define CLKRC	0x11
#define CLKRC_VALUE_VGA	0x01 // Raw Bayer
#define CLKRC_VALUE_QVGA	0x01
#define CLKRC_VALUE_QQVGA	0x01
#define CLKRC_VALUE_NIGHTMODE_FIXED	0x03 // Fixed Frame
#define CLKRC_VALUE_NIGHTMODE_AUTO	0x80 // Auto Frame Rate Adjust
#define COM7	0x12
#define COM7_VALUE_VGA	0x01 // Raw Bayer
#define COM7_VALUE_VGA_COLOR_BAR	0x03 // Raw Bayer
#define COM7_VALUE_VGA_PROCESSED_BAYER	0x05 // Processed Bayer
#define COM7_VALUE_QVGA	0x00
#define COM7_VALUE_QVGA_COLOR_BAR	0x02
#define COM7_VALUE_QVGA_PREDEFINED_COLOR_BAR	0x12
#define COM7_VALUE_QQVGA	0x00
#define COM7_VALUE_QQVGA_COLOR_BAR	0x02
#define COM7_VALUE_RESET	0x80
#define COM3	0x0C
#define COM3_VALUE_VGA	0x00 // Raw Bayer
#define COM3_VALUE_QVGA	0x04

```c
#define COM3_VALUE_QQVGA                    0x04 // From Docs
#define COM3_VALUE_QQVGA_SCALE_ENABLED      0x0C // Enable Scale and DCW


#define COM14                               0x3E
#define COM14_VALUE_VGA                     0x00 // Raw Bayer
#define COM14_VALUE_QVGA                    0x19
#define COM14_VALUE_QQVGA                   0x1A
#define COM14_VALUE_MANUAL_SCALING          0x08 // Manual Scaling Enabled
#define COM14_VALUE_NO_MANUAL_SCALING       0x00 // Manual Scaling DisEnabled


#define SCALING_XSC                         0x70
#define SCALING_XSC_VALUE_VGA               0x3A // Raw Bayer
#define SCALING_XSC_VALUE_QVGA              0x3A
#define SCALING_XSC_VALUE_QQVGA             0x3A
#define SCALING_XSC_VALUE_QQVGA_SHIFT1      0x3A
#define SCALING_XSC_VALUE_COLOR_BAR         0xBA


#define SCALING_YSC                         0x71
#define SCALING_YSC_VALUE_VGA               0x35 // Raw Bayer
#define SCALING_YSC_VALUE_QVGA              0x35
#define SCALING_YSC_VALUE_QQVGA             0x35
#define SCALING_YSC_VALUE_COLOR_BAR         0x35 // 8 bar color bar
#define SCALING_YSC_VALUE_COLOR_BAR_GREY    0xB5 // fade to grey color bar
#define SCALING_YSC_VALUE_COLOR_BAR_SHIFT1  0xB5 // fade to grey color bar


#define SCALING_DCWCTR                      0x72
#define SCALING_DCWCTR_VALUE_VGA            0x11 // Raw Bayer
#define SCALING_DCWCTR_VALUE_QVGA           0x11
#define SCALING_DCWCTR_VALUE_QQVGA          0x22
```

```c
#define SCALING_PCLK_DIV                                0x73
#define SCALING_PCLK_DIV_VALUE_VGA                      0xF0 // Raw Bayer
#define SCALING_PCLK_DIV_VALUE_QVGA                     0xF1
#define SCALING_PCLK_DIV_VALUE_QQVGA                    0xF2


#define SCALING_PCLK_DELAY                              0xA2
#define SCALING_PCLK_DELAY_VALUE_VGA                    0x02 // Raw Bayer
#define SCALING_PCLK_DELAY_VALUE_QVGA                   0x02
#define SCALING_PCLK_DELAY_VALUE_QQVGA                  0x02


// Controls YUV order Used with COM13
// Need YUYV format for Android Decoding- Default value is 0xD

#define TSLB                                            0x3A
#define TSLB_VALUE_YUYV_AUTO_OUTPUT_WINDOW_ENABLED      0x01 // No custom scaling
#define TSLB_VALUE_YUYV_AUTO_OUTPUT_WINDOW_DISABLED     0x00 // For adjusting HSTART, etc. YUYV format
#define TSLB_VALUE_UYVY_AUTO_OUTPUT_WINDOW_DISABLED     0x08
#define TSLB_VALUE_TESTVALUE                            0x04 // From YCbCr Reference


// Default value is 0x88
// ok if you want YUYV order, no need to change

#define COM13                                           0x3D
#define COM13_VALUE_DEFAULT                             0x88
#define COM13_VALUE_NOGAMMA_YUYV                        0x00
#define COM13_VALUE_GAMMA_YUYV                          0x80
#define COM13_VALUE_GAMMA_YVYU                          0x82
#define COM13_VALUE_YUYV_UVSATAUTOADJ_ON                0x40
```

```c
// Works with COM4
#define COM17                                        0x42
#define COM17_VALUE_AEC_NORMAL_NO_COLOR_BAR          0x00
#define COM17_VALUE_AEC_NORMAL_COLOR_BAR             0x08 // Activate Color Bar for DSP


#define COM4   0x0D


// RGB Settings and Data format
#define COM15   0x40


// Night Mode
#define COM11                                    0x3B
#define COM11_VALUE_NIGHTMODE_ON                 0x80   // Night Mode
#define COM11_VALUE_NIGHTMODE_OFF                0x00
#define COM11_VALUE_NIGHTMODE_ON_EIGHTH          0xE0   // Night Mode 1/8 frame rate minimum
#define COM11_VALUE_NIGHTMODE_FIXED              0x0A
#define COM11_VALUE_NIGHTMODE_AUTO               0xEA   // Night Mode Auto Frame Rate Adjust


// Color Matrix Control YUV
#define MTX1              0x4f
#define MTX1_VALUE        0x80


#define MTX2              0x50
#define MTX2_VALUE        0x80


#define MTX3              0x51
#define MTX3_VALUE        0x00
```

```c
#define MTX4              0x52
#define MTX4_VALUE        0x22

#define MTX5              0x53
#define MTX5_VALUE        0x5e

#define MTX6              0x54
#define MTX6_VALUE        0x80

#define CONTRAS           0x56
#define CONTRAS_VALUE     0x40

#define MTXS              0x58
#define MTXS_VALUE        0x9e

// COM8
#define COM8                      0x13
#define COM8_VALUE_AWB_OFF        0xE5
#define COM8_VALUE_AWB_ON         0xE7

// Automatic White Balance
#define AWBC1             0x43
#define AWBC1_VALUE       0x14

#define AWBC2             0x44
#define AWBC2_VALUE       0xf0

#define AWBC3             0x45
#define AWBC3_VALUE       0x34
```

```c
#define AWBC4            0x46
#define AWBC4_VALUE      0x58


#define AWBC5            0x47
#define AWBC5_VALUE      0x28


#define AWBC6            0x48
#define AWBC6_VALUE      0x3a


#define AWBC7            0x59
#define AWBC7_VALUE      0x88


#define AWBC8            0x5A
#define AWBC8_VALUE      0x88


#define AWBC9            0x5B
#define AWBC9_VALUE      0x44


#define AWBC10           0x5C
#define AWBC10_VALUE     0x67


#define AWBC11           0x5D
#define AWBC11_VALUE     0x49


#define AWBC12           0x5E
#define AWBC12_VALUE     0x0E


#define AWBCTR3                  0x6C
```

```c
#define AWBCTR3_VALUE            0x0A


#define AWBCTR2                  0x6D
#define AWBCTR2_VALUE            0x55


#define AWBCTR1                  0x6E
#define AWBCTR1_VALUE            0x11


#define AWBCTR0                  0x6F
#define AWBCTR0_VALUE_NORMAL     0x9F
#define AWBCTR0_VALUE_ADVANCED   0x9E


// Gain
#define COM9                     0x14
#define COM9_VALUE_MAX_GAIN_128X 0x6A
#define COM9_VALUE_4XGAIN        0x10   // 0001 0000


#define BLUE         0x01  // AWB Blue Channel Gain
#define BLUE_VALUE   0x40


#define RED          0x02  // AWB Red Channel Gain
#define RED_VALUE    0x40


#define GGAIN        0x6A  // AWB Green Channel Gain
#define GGAIN_VALUE  0x40


#define COM16        0x41
#define COM16_VALUE  0x08 // AWB Gain on
```

```c
#define GFIX                0x69
#define GFIX_VALUE          0x00


// Edge Enhancement Adjustment
#define EDGE                0x3f
#define EDGE_VALUE          0x00


#define REG75               0x75
#define REG75_VALUE         0x03


#define REG76               0x76
#define REG76_VALUE         0xe1


// DeNoise
#define DNSTH               0x4c
#define DNSTH_VALUE         0x00


#define REG77               0x77
#define REG77_VALUE         0x00


// Denoise and Edge Enhancement
#define COM16_VALUE_DENOISE_OFF_EDGE_ENHANCEMENT_OFF_AWBGAIN_ON    0x08 // Denoise off, AWB Gain on
#define COM16_VALUE_DENOISE_ON__EDGE_ENHANCEMENT_OFF__AWBGAIN_ON   0x18
#define COM16_VALUE_DENOISE_OFF__EDGE_ENHANCEMENT_ON__AWBGAIN_ON   0x28
#define COM16_VALUE_DENOISE_ON__EDGE_ENHANCEMENT_ON__AWBGAIN_ON    0x38 // Denoise on, Edge Enhancement on, AWB Gain on


// 30FPS Frame Rate , PCLK = 24Mhz
```

```
#define CLKRC_VALUE_30FPS  0x80


#define DBLV                    0x6b
#define DBLV_VALUE_30FPS        0x0A


#define EXHCH                   0x2A
#define EXHCH_VALUE_30FPS       0x00


#define EXHCL                   0x2B
#define EXHCL_VALUE_30FPS       0x00


#define DM_LNL                  0x92
#define DM_LNL_VALUE_30FPS      0x00


#define DM_LNH                  0x93
#define DM_LNH_VALUE_30FPS      0x00


#define COM11_VALUE_30FPS       0x0A


// Saturation Control
#define SATCTR          0xc9
#define SATCTR_VALUE    0x60


// AEC/AGC - Automatic Exposure/Gain Control
#define GAIN            0x00
#define GAIN_VALUE      0x00


#define AEW             0x24
#define AEW_VALUE       0x95
```

```
#define AEB                             0x25
#define AEB_VALUE                       0x33


#define VPT                             0x26
#define VPT_VALUE                       0xe3


// AEC/AGC Control- Histogram
#define HAECC1                          0x9f
#define HAECC1_VALUE                    0x78


#define HAECC2                          0xa0
#define HAECC2_VALUE                    0x68


#define HAECC3                          0xa6
#define HAECC3_VALUE                    0xd8


#define HAECC4                          0xa7
#define HAECC4_VALUE                    0xd8


#define HAECC5                          0xa8
#define HAECC5_VALUE                    0xf0


#define HAECC6                          0xa9
#define HAECC6_VALUE                    0x90


#define HAECC7                          0xaa  // AEC Algorithm selection
#define HAECC7_VALUE_HISTOGRAM_AEC_ON   0x94
#define HAECC7_VALUE_AVERAGE_AEC_ON     0x00
```

```c
// Array Control
#define CHLF              0x33
#define CHLF_VALUE        0x0b

#define ARBLM             0x34
#define ARBLM_VALUE       0x11

// ADC Control
#define ADCCTR1           0x21
#define ADCCTR1_VALUE     0x02

#define ADCCTR2           0x22
#define ADCCTR2_VALUE     0x91

#define ADC               0x37
#define ADC_VALUE         0x1d

#define ACOM              0x38
#define ACOM_VALUE        0x71

#define OFON              0x39
#define OFON_VALUE        0x2a

// Black Level Calibration
#define ABLC1             0xb1
#define ABLC1_VALUE       0x0c

#define THL_ST            0xb3
```

```c
#define THL_ST_VALUE     0x82


// Window Output

#define HSTART                       0x17
#define HSTART_VALUE_DEFAULT         0x11
#define HSTART_VALUE_VGA             0x13
#define HSTART_VALUE_QVGA            0x13
#define HSTART_VALUE_QQVGA           0x13   // Works


#define HSTOP                        0x18
#define HSTOP_VALUE_DEFAULT          0x61
#define HSTOP_VALUE_VGA              0x01
#define HSTOP_VALUE_QVGA             0x01
#define HSTOP_VALUE_QQVGA            0x01   // Works


#define HREF                         0x32
#define HREF_VALUE_DEFAULT           0x80
#define HREF_VALUE_VGA               0xB6
#define HREF_VALUE_QVGA              0x24
#define HREF_VALUE_QQVGA             0xA4


#define VSTRT                        0x19
#define VSTRT_VALUE_DEFAULT          0x03
#define VSTRT_VALUE_VGA              0x02
#define VSTRT_VALUE_QVGA             0x02
#define VSTRT_VALUE_QQVGA            0x02


#define VSTOP                        0x1A
#define VSTOP_VALUE_DEFAULT          0x7B
```

```c
#define VSTOP_VALUE_VGA        0x7A
#define VSTOP_VALUE_QVGA       0x7A
#define VSTOP_VALUE_QQVGA      0x7A


#define VREF                   0x03
#define VREF_VALUE_DEFAULT     0x03
#define VREF_VALUE_VGA         0x0A
#define VREF_VALUE_QVGA        0x0A
#define VREF_VALUE_QQVGA       0x0A
```

Appendix B: Image Capture Program Variables

```
const int chipSelect = 48;

const int HardwareSSPin = 53;        // For Arduino Mega


int PhotoTakenCount = 0;


// Serial Input

const int BUFFERLENGTH = 255;

char IncomingByte[BUFFERLENGTH];    // for incoming serial data


// VGA Default

int PHOTO_WIDTH  =  640;

int PHOTO_HEIGHT =  480;

int PHOTO_BYTES_PER_PIXEL = 2;


// Command and Parameter related Strings

String RawCommandLine = "";

String Command  = "QQVGA";

String FPSParam = "ThirtyFPS";

String AWBParam = "SAWB";

String AECParam = "HistAEC";

String YUVMatrixParam = "YUVMatrixOn";

String DenoiseParam = "DenoiseNo";
```

```cpp
String EdgeParam = "EdgeNo";

String ABLCParam = "AblcON";


enum ResolutionType
{
  VGA,
  VGAP,
  QVGA,
  QQVGA,
  None
};

ResolutionType Resolution = None;


// Camera input/output pin connection to Arduino

#define WRST  22        // Output Write Pointer Reset
#define RRST  23        // Output Read Pointer Reset
#define WEN   24        // Output Write Enable
#define VSYNC 25        // Input Vertical Sync marking frame capture
#define RCLK  26        // Output FIFO buffer output clock
// set OE to low gnd


// FIFO Ram input pins
#define DO7   28
#define DO6   29
#define DO5   30
#define DO4   31
#define DO3   32
#define DO2   33
#define DO1   34
```

```c
#define DO0   35


// I2C

#define OV7670_I2C_ADDRESS                  0x21
#define I2C_ERROR_WRITING_START_ADDRESS     11
#define I2C_ERROR_WRITING_DATA              22


#define DATA_TOO_LONG                 1   // data too long to fit in transmit buffer
#define NACK_ON_TRANSMIT_OF_ADDRESS   2   // received NACK on transmit of address
#define NACK_ON_TRANSMIT_OF_DATA      3   // received NACK on transmit of data
#define OTHER_ERROR                   4   // other error


#define I2C_READ_START_ADDRESS_ERROR           33
#define I2C_READ_DATA_SIZE_MISMATCH_ERROR      44
```

www.ingramcontent.com/pod-product-compliance
Lightning Source LLC
Chambersburg PA
CBHW080803180526
45168CB00006B/2313